ENVIRONMENTAL SCIENCE, ENGINEERING AND TECHNOLOGY

THE INFLUENCE OF ECOSYSTEM SERVICES TOWARDS HUMAN WELLBEING

Environmental Science, Engineering and Technology

Additional books and e-books in this series can be found on Nova's website under the Series tab.

ENVIRONMENTAL SCIENCE, ENGINEERING AND TECHNOLOGY

THE INFLUENCE OF ECOSYSTEM SERVICES TOWARDS HUMAN WELLBEING

HASMAH ABDULLAH
AND
RAPEAH SUPPIAN
EDITORS

Library of Congress Cataloging-in-Publication Data

ISBN: 978-1-53619-977-2

Published by Nova Science Publishers, Inc. † New York

CONTENTS

PREFACE

Ecosystem services provide benefits to humans, including provisioning services (food, water, timber, fibre and genetic resources), regulating services (regulation of climate, floods, diseases and water quality), cultural services (recreational, aesthetic and spiritual), and support services (soil formation, pollination and nutrient cycling). Promoting the concept of ecosystem services reveals the potential of its contribution to environmental wellbeing for conservation and sustainability. Humans, as users of the provided services, benefit from ecosystem services, fostering dependency on nature. The use of ecosystem services may lead to unintended environmental consequences throughout the supply chain. Hence, this book will focus on the services provided for human wellbeing and a multilayer association with human problems worldwide when supply chains are disturbed.

Carbon emissions from city areas are responsible for 75% of world carbon dioxide emissions, making them a significant contributor to climate change. Urban populations are early responders to the impacts of climate change. In addition, ecosystem services are influenced by human activities. Domestic and industrial water pollution has led to the contamination of drinking water in many parts of the world. Waste management, such as solid waste management and biomedical waste management, is considered crucial for balancing ecosystem services as a consequence of human activities. In addition, air pollution has become part of ecosystem service disturbance, as it has become a worldwide problem and a major threat to the surrounding environment and human health. The major sources of air pollutants are mobile sources and stationary and transboundary emissions. Human activities, such as mining and exploration, have brought naturally occurring radioactive elements, such as gamma rays, which are present at relatively low concentrations in many geologic formations and earth materials, to the surface and have become a threat to human health. Among others, loud noise exposure in occupational settings has been found to be hazardous to hearing organs.

Environmental stressors that lead to ecosystem changes have been shown to trigger noncommunicable diseases, such as cardiovascular disease and cancer. The emergence and re-emergence of vector-borne diseases, such as Zika, Dengue and Malaria, are rapidly influenced by changes in ecosystem services. Many studies have been conducted to develop effective vaccines to combat such diseases. New drugs have been developed from natural resources with the aim of combating drug resistance, enhancing efficacy and reducing toxicity. In addition, a health education programme (HEP) is also being developed to improve the quality of life of patients. A broad spectrum of research is reported in this book covering environmental monitoring, modelling, molecular research, natural product discovery and health education programmes, showing the importance of support from ecosystem services that must be preserved for future wellbeing.

Chapter 1 - Individuals who are exposed to biomedical waste (BMW) are at danger and, if not treated appropriately, can cause environmental deterioration. To avoid mishaps, every laboratory technologist is expected to have a high level of knowledge, attitude, and practise (KAP). The aim of this study is to assess the levels of knowledge, attitude, and practices among laboratory technologists in selected departments at Hospital Universiti Sains Malaysia (HUSM), and to find the association of demographic data with those three domains' levels, then to find the relationship with KAP and compare KAP score on BMW management among laboratory technologists in those departments. A self-administered questionnaire has been used to conduct a cross-sectional study on BMW management knowledge, attitude, and practise. Laboratory technologists at Pathology, Medical Tranfusion Unit (UPT) and Haematology, Microbiology and Parasitology, and Emergency departments were among 74 participants. This study resulted in a moderate level of knowledge among all laboratory technologists who participated. While, high level of attitude and practice was found amongst laboratory technologists. While a high level of attitude and practice was found amongst laboratory technologists. However, an association was found between departments with the level of attitude toward BMW management ($p = 0.024$). While, knowledge and practice scores were correlated with weak and positive correlation ($r = 0.255$, $p = 0.029$). Lastly, there was a significant difference in attitude scores between selected departments ($p < 0.001$). When compared to other departments, UPT and the Haematology department have much higher attitude scores (6, IQR 1). The knowledge, attitude, and practise of laboratory technologists on BMW management can change as a result of greater understanding and positive actions toward waste disposal, which can lead to improved BMW management practise.

Chapter 2 - Cities are responsible for 75% of global carbon dioxide emissions, making them a major contributor to climate change (CC). Since the urban population is the first to experience the impacts of climate change, their knowledge, attitude, and practise (KAP) were assessed. A self-administered questionnaire has been used to conduct a cross-sectional analysis of 232 respondents aged 15 and above. Five out of eleven districts in Kuala Lumpur (KL) were selected by simple random sampling, while subjects were recruited from community settings using convenient sampling. The relationship between KAP scores was investigated using Spearman correlation tests. To determine t, the Chi-Square test was used. There was an association between knowledge and awareness ($r = 0.302$), knowledge and practise ($r = 0.142$) and awareness and practise ($r = 0.196$). The educational level had a significant association with knowledge and awareness. The community had knowledge about CC, however, they are not aware of what actions to be taken. The government shall disseminate information about CC through mass and social media. Combinations of top-down and bottom-up approaches are important for adaptation and mitigation strategies.

Chapter 3 - The Zika virus is a vector-borne disease that shares the same transmission mode as dengue fever. It has the potential to cause neurological problems in children. Several cases of Zika disease had been recorded in Malaysia before December 2016. Then, in order to respond to epidemics and pandemics, it was necessary to determine awareness, attitude, and practise. The authors' aim was to investigate Zika disease awareness, attitudes, and practises among students of Environmental and Occupational Health at a Malaysian public university. A cross-sectional survey and self-administered questionnaire was distributed using universal sampling. Purposive sampling was then used to choose the stage of study year. The knowledge, attitude and practise on Zika disease were measured using adapted survey tools by World Health Organization. Descriptive analysis showed the level of knowledge, attitude and practise of 172 students Environmental and Occupational Health were at a moderate level. However, the result of an association between socio-demographic and practise, showed a significant difference ($p = 0.005$) between age and practise towards prevention of Zika disease. Meanwhile, a correlation test was conducted to identify if there was any correlation between knowledge, attitude and practise. Then, there was a weak significant positive correlation between knowledge and attitude ($r = 0.205$, $p = 0.007$) and between practise and attitude ($r = 0.549$, $p = 0.001$). The result also had shown no significant difference in knowledge, attitude and practise between first and fourth-year students. By determining the level of knowledge, attitude and practise of Zika disease among university students, it can provide baseline data and guides to health authorities in order to plan preventive measure programs in future.

Chapter 4 - Roundabouts are designed to replace traffic signalised intersections because they are thought to improve traffic flow, minimise fuel consumption, and reduce noise levels. It is, however, vulnerable to a variety of air pollutants, and one of those is particulate matter. The objective of this research is to determine the changes of selected particulate matter ($PM_{2.5}$ and PM_{10}) concentrations caused by vehicular emissions and their impacts on urban air quality in Kelantan, as well as to compare $PM_{2.5}$ and PM_{10} concentrations between different roundabouts and to determine the relationship between $PM_{2.5}$ and PM_{10} concentrations and vehicle count. $PM_{2.5}$ and PM_{10} concentrations were measured at three selected roundabouts in Kelantan by using DustTrak DRX Desktop 8533. The equipment was located at the centre of the roundabout for two hours per session. Besides, a vehicle counter was used to count vehicles manually according to their type. Data collection was done for seven days continuously at each roundabout. The highest readings were recorded at the roundabout at Pasir Pekan, (336 ± 69.66 μg/m^3) for $PM_{2.5}$ and (352 ± 73.41 μg/m^3) for PM10. There was no significant difference of PM2.5 ($p = 0.345$) and PM_{10} ($p = 0.238$) concentrations between three selected roundabouts. However, there was a significant correlation between the number of vehicles and particulate matter concentrations, which are $PM_{2.5}$ ($p = 0.003$) and PM_{10} ($p = 0.004$). Roundabout at Pasir Pekan is considered the most polluted, as it has the highest mean and median of $PM_{2.5}$ and PM_{10} concentrations compared to the other two roundabouts.

Chapter 5 - Heavy metal pollution of wastewater is a serious issue for both humans and the environment. The concentration of heavy metals such as lead (Pb), cadmium (Cd), chromium (Cr), zinc (Zc), and copper (Cu) in wastewater discharged from industrial, agriculture, and domestic household areas were investigated in this study. The elements were analysed using atomic absorption spectroscopy (AAS) after the wastewater sample was digested using the wet acid digestion method. The highest concentration of Cd and Cu were found in the industrial area, while the highest concentration of Pb, Zn, and Cr were detected in the agricultural farm area. Cd effluent in wastewater of industrial sites, agriculture farms and domestic households found to be exceeded the permissible level. These results revealed that proper wastewater treatment and monitoring are vitally crucial for the three areas.

Chapter 6 - Measurements for Terrestrial Gamma Radiation (TGR) dose rate were made in Michika area of north-eastern Nigeria using aeromagnetic survey in 6983 locations with a range value from 15 nGy h-1 to 324 nGy h^{-1} and the mean value of 115 ± 0.5 nGy h^{-1}. The outdoor annual effective dose is found to be 0.141 mSv $y.^{-1}$. The computed lifetime effective dose, cancer risk and the lifetime cancer risk from outdoors exposure for each person living in the Michika area is found to be 7.614 mSv, 8.21×10^{-3} and 4.43×10^{-1} respectively. These values are two times higher than the world outdoor average values of 4.9 mSv, 4.07×10^{-3}, and 2.85×10^{-1} respectively. The effective doses are possibly due to inhalation of gamma radiation targeting the internal organs like the gonads (testes or ovaries), lung, liver, and the skin of the body with values of 0.023μSv, 0.014μSv, 0.006μSv and 0.001μSv respectively. An Isodose map of the study area was presented to depict areas of enhancing TGR.

Chapter 7 - Smoking is one of the most important public health issues among communities which may lead to health implications. Owing to this fact, this study was conducted to investigate the effects of smoking and secondhand smoke (SHS) exposure at home on the self-reported respiratory symptoms among adults in Kg. Beris Lalang, Bachok, Kelantan. A cross-sectional study among 218 respondents aged 18 - 65 years old was conducted using a purposive sampling through door to door survey. The socio-demographic variables examined were age, gender, occupation and level of education. The prevalence of smokers was associated with five respiratory symptoms including wheezing (46.8%), wheezing without cold (50.9%), chronic cough (63.2%), phlegm (36.7%) and sore throat (61.1%). Meanwhile, secondhand smoke exposure was associated with more prevalence in chest tightness (43.6%), coughing attack (43.2%), breathing trouble (46.7%) and sneezing, runny or blocked nose (40.5%) compared to non-smokers and non-exposed respondents (Chi-square test, $p < 0.001$). This study highlights the importance of minimising smoke exposure at home due to the prevalence of both active and passive smokers to develop respiratory symptoms related to tobacco smoke.

Chapter 8 - Emergency responders have a higher risk of injury and death than other professions. Multiple factors should be identified because they are most vulnerable to depression, anxiety, and stress, which could have a negative impact on health. However, no such research has been conducted on Fire and Rescue Department Malaysia (FRDM) responders.

Therefore, this study was conducted to assess depression, anxiety, and stress among FRDM responders, as well as their relationship to the factors that contribute to it. A questionnaire from DASS-21 and NIOSH-GJSQ was used to assess depression, anxiety, stress level, and its variables in 47 respondents from FRDM Kajang Station, Selangor. The finding showed that, 40.4% of respondents reported higher-than-average levels of depression, anxiety, and stress, while general health, social support from supervisors, and job satisfaction factors identified as contributing factors. In conclusion, social support from a co-worker as well as social support from family and friends may help respondents manage their depression, anxiety, and stress.

Chapter 9 - Composting is the natural recycling of organic matter. The process takes several months to produce a high-quality mature compost that can be used as a fertilizer. This study was designed to track the time it takes for different types of compost to mature and to compare the growth of *Ipomoea aquatica* after 21 days of planting with different types of compost. The levels of total nitrogen, total phosphorus, and potassium in the various types of compost were also compared. Organic waste was collected from the cafeteria at the Universiti Sains Malaysia Health Campus. Fruit peels, food leftovers, and mixed organic waste were allowed to decompose for 60 to 90 days. The mature compost was then used as a planting medium for *Ipomoea aquatica*. Acid digestion was used for macro element analysis. The potassium level was determined using atomic absorption spectroscopy (AAS), while total nitrogen and total phosphorus were determined using a COD Reactor DRB200 and a colorimeter (Hach Model DR/890). The results showed that the time required to mature compost from fruit peels was 60 days, while the others took 75 days. Furthermore, there was a significant difference in *Ipomoea aquatica* growth when different types of organic waste compost were used. There was also a significant difference in macro elements between the various types of organic waste compost. These findings indicate that composts made from various organic wastes differed in terms of quality and stability, which is further influenced by the composition of the raw material used in compost production.

Chapter 10 - This study predicts the three-dimensional structure of chitin deacetylase using the method of homology modeling. The objective of this study is to compare the predicted structure with the native experimentally solved structure. This method involves the identification of template protein, alignment of the target sequence to the template, generation of the 3D model, and model evaluation. In addition to that, the location of the catalytic sites was predicted as well. The accuracy of this method was highly dependent on the percentage of similarity between the target and the template protein. In conclusion, the computational method complements the experimental structure in predicting protein structures.

Chapter 11 - A recombinant BCG vaccine expressing the synthetic MSP-1C of *Plasmodium falciparum* was developed in the previous study. The vaccine candidate increased humoral and cellular immune responses in mice, as well as triggering phagocytosis activity and producing pro-inflammatory cytokines that were significantly higher than those produced by the parent BCG clone. However, no research has been done to assess the stability of the vaccine candidate. The acidity or alkalinity of the growth medium is one of the most important factors affecting vaccine stability. The purpose of this study was to determine the effect of pH on the stability of the rBCG candidate vaccine. For 14 days, the rBCG culture and parent BCG control were grown in different pH levels of 7H9 broth (pH 5, pH 6, pH 7, and pH 9). On day 14, the viability of the mycobacteria was determined using a spectrophotometer with an optical density (OD) of 600 nm. The effect of pH on the morphology of the mycobacteria was determined using scanning electron microscope while the effect of pH on the stability of MSP-1C gene in the rBCG clone was determined using polymerase chain reaction. The results showed that cloning the MSP-1C gene into the BCG genome does not affect the growth, morphology, or stability of the rBCG vaccine. Both rBCG and the parent BCG control have a similar response to different pH levels of 7H9 broth. The optimum pH for the growth and morphological stability of the rBCG and parent BCG, as well as the stability of the MSP-1C gene in the rBCG clone was at pH 6 - pH 7 respectively, while the growth of the mycobacteria was significantly reduced at pH 5 and pH 9. The results also showed that all the rBCG clones from different pH of 7H9 expressed MSP-1C gene indicating the stability of the gene. Accordingly, these findings paved the way for the clinical application of rBCG as a vaccine against malaria in the future.

In: The Influence of Ecosystem Services … ISBN: 978-1-53619-977-2
Editors: Hasmah Abdullah et al.

Chapter 1

KNOWLEDGE, ATTITUDE AND PRACTICE OF BIOMEDICAL WASTE MANAGEMENT AMONG LABORATORY TECHNOLOGISTS OF SELECTED DEPARTMENTS IN THE HOSPITAL UNIVERSITI SAINS MALAYSIA

Nur Fatihah R., Nurulilyana S.*, Nurkhairina Syahirah S. and Raja Nur Shafieza R. M.
School of Health Sciences, Health Campus,
Universiti Sains Malaysia, Kubang Kerian, Kelantan, Malaysia

ABSTRACT

Individuals who are exposed to biomedical waste (BMW) are at danger and, if not treated appropriately, can cause environmental

---

* Corresponding Author's E-mail: nurulilyana@usm.my.

deterioration. To avoid mishaps, every laboratory technologist is expected to have a high level of knowledge, attitude, and practise (KAP). The aim of this study is to assess the levels of knowledge, attitude, and practices among laboratory technologists in selected departments at Hospital Universiti Sains Malaysia (HUSM), and to find the association of demographic data with those three domains' levels, then to find the relationship with KAP and compare KAP score on BMW management among laboratory technologists in those departments. A self-administered questionnaire has been used to conduct a cross-sectional study on BMW management knowledge, attitude, and practise. Laboratory technologists at Pathology, Medical Tranfusion Unit (UPT) and Haematology, Microbiology and Parasitology, and Emergency departments were among 74 participants. This study resulted in a moderate level of knowledge among all laboratory technologists who participated. While, high level of attitude and practice was found amongst laboratory technologists. While a high level of attitude and practice was found amongst laboratory technologists. However, an association was found between departments with the level of attitude toward BMW management ($p = 0.024$). While, knowledge and practice scores were correlated with weak and positive correlation ($r = 0.255$, $p = 0.029$). Lastly, there was a significant difference in attitude scores between selected departments ($p < 0.001$). When compared to other departments, UPT and the Haematology department have much higher attitude scores (6, IQR 1). The knowledge, attitude, and practise of laboratory technologists on BMW management can change as a result of greater understanding and positive actions toward waste disposal, which can lead to improved BMW management practise.

Keywords: BMW management, knowledge, attitude, practice, laboratory technologist

INTRODUCTION

Biomedical waste management (BMW) is any unwanted product generated during diagnosis, treatment and immunization of animals and humans that includes research activities (Sachan et al. 2012). From the Ministry of Environment and Forest Notification on Biomedical Waste (Management and Handling), Rules 1998 stated that all healthcare persons need to ensure the segregation, safe handling and disposal of BMW without causing any adverse effects to human health and the environment. BMW has a higher risk than other types of waste (Park 2011).

Annually, approximately 0.33 million tons of hospital waste is generated in India, and the waste generation rate ranges from 0.5 kg to 2.0 kg per day (Patil and Shekdar 2001). In addition, it is estimated that 10-25% of healthcare waste is hazardous and has the potential to create a variety of health problems for human health and the environment (Yadannavar et al. 2010). For instance, human immunodeficiency virus (HIV) as well as hepatitis B and C or any other infectious disease are also concerns caused by needle-stick injuries and other forms of contagion (Sachan et al. 2012). The World Health Organization (WHO) reported that approximately 85% of hospital waste is actually non-hazardous, 10% is hazardous waste and the remaining 5% is non-infectious. However, BMW can be in the form of any solid, fluid or liquid waste, including its container and any intermediate product (Santpathy and Pandhi 1998).

According to the Department of Environment (DoE), Malaysia (2009), it is estimated that the rate of generation for clinical waste ranges from 0.3 to 0.8 kg per occupied bed per day. In Malaysia, the DoE is accounted for under the Environmental Quality Act 1974 to control and prevent pollution as well as to protect and enhance the quality of the environment, and clinical waste is classified as scheduled waste under the Environmental Quality (Scheduled Wastes) Regulations, 2005. Hence, adequate knowledge of health hazards, proper techniques, methods of handling waste and safety practices can lead to safe BMW management in addition to preventing the community from experiencing various adverse effects (Madhukumar and Ramesh 2012).

METHODOLOGY

Study Design and Study Population

A cross-sectional study was carried out in a ten-month period (from September 2018 until June 2019) involving all 88 laboratory technologists working in the Pathology, Medical Tranfusion Unit (UPT) and Haematology, Microbiology and Parasitology, and Emergency departments in the Hospital Universiti Sains Malaysia (HUSM), Kubang Kerian, Kelantan. The study protocol was approved by the Human Research Ethical Committee (HREC) of Universiti Sains Malaysia.

Sampling Method

Purposive sampling was used to select participants in this study. Participants were chosen in every selected department based on the inclusion criteria. Subjects who were laboratory technologists who had handled BMW for more than six months were selected for this study. Then, the questionnaires were given to the person in charge of the laboratory technologist at each department for distribution. The questionnaires were left for a week for the participants to complete. It took fifteen to twenty minutes to answer all the questions. During data collection, only 74 participants met the inclusion criteria and agreed to participate in this study.

Research Tools

A self-administered questionnaire was used to collect data in this study. This questionnaire was a modified version from a previous study after acquiring permission from Sanjeev et al. (2014) and Malini and Eshwar (2015). The questionnaire was translated into Malay language. It consists of four sections: section A relates to the demographic data of the participant, and sections B, C and D relate to the knowledge, attitudes and practices regarding BMW management. Knowledge includes nine questions, attitudes contains eight questions and practices contains 12 questions. The questionnaire was pretested by conducting a pilot study (n = 9) to ensure that the questionnaire was reliable, valid and understood by participants. The scoring was as follows: '1' for a correct score and '0' for a wrong one. The grading of the scores was analysed using the sum score of each outcome based on Bloom's cut-off point (60-80%).

Data Analysis

The data were analysed using the Statistical Package for Social Science (SPSS) version 24.0. Descriptive statistics were used to calculate the frequency distribution and percentage for data collection. In addition, inferential statistics were used to test the hypotheses. Pearson's chi-square test was used to examine the association of demographic data with KAP, and Spearman's correlation test was used to determine the correlations between knowledge, attitude and practice. Finally, the Kruskal-Wallis test was used to compare KAP scores between selected departments.

Ethical Consideration

Approval from participants was obtained through an information sheet and consent form. Permission from the Head of HUSM and each head of the department was obtained to conduct this study. Meanwhile, ethical approval from the Human Research Ethics Committee (HREC) USM through the Research Ethics Approval Application form was obtained before starting this study.

Results

Demographic Data of Participant

Table 1 shows that 74 participants agreed to voluntarily participate in this study. The majority of them were female, aged between 31 and 40 years, Malay, diploma holders and had working experience of approximately 11 to 20 years.

Table 1. Demographic data of participants

Categories		**N (%)**
Gender	Female	46 (62.2)
Age (years)	31-40	45 (60.8)
Race	Malay	71 (95.9)
Education level	Diploma	54 (73.0)
Work (years)	11-20	33 (44.6)
Pathology department		14 (18.9)
UPT & haematology department		39 (52.7)
Microbiology & parasitology department		19 (25.7)
Emergency department		2 (2.7)

Table 2. Descriptive results of knowledge on BMW management

Knowledge	**N (%)**
Are all healthcare wastes hazardous	66 (89.2)
Are you aware that biomedical waste management rules exist in your department	71 (95.9)
Can ordinary plastic bags be used for waste disposal	72 (97.3)
Have you had training in biomedical waste management	42 (56.8)
Are you aware of IMAGE	5 (6.8)
What does IMAGE stand for	5 (6.8)
Do you know the maximum time limit for biomedical waste that can be stored	11 (14.9)
What is the maximum time limit during which biomedical waste can be stored	2 (2.7)
Which of the following is the universally accepted symbol for biohazard	70 (94.6)

Knowledge on BMW Management

Table 2 shows that the majority of the participants had a moderate level of knowledge: 78.6% Pathology, 79.5% UPT and Haematology, 94.7% Microbiology and Parasitology and 100.0% Emergency. Table 2 shows the percentage of correctly answered questions on knowledge. Approximately 89.2% of the participants considered all healthcare waste hazardous. Most of them were aware of the fact that BMW management and handling rules were applicable to their department. Approximately 97.3% knew that ordinary plastic bags cannot be used for waste disposal. Approximately half of the participants had received training on BMW management. Surprisingly, only 6.8% were aware of IMAGE and knew the correct expansion of the abbreviation IMAGE. However, only 2.7% knew the maximum storage period for the BMW was 48 hours. Almost 94.6% correctly recognized the symbol of biohazard.

Attitudes on BMW Management

Table 3 shows that most participants had a strong attitude regarding BMW management: 78.6% Pathology, 84.6% UPT and Haematology, 57.9% Microbiology and Parasitology and 100.0% Emergency. Table 3 shows that a total of 97.3% of them agreed that BMW should be segregated into different categories.

Table 3. Descriptive results of attitudes on BMW management

Attitude	N (%)
Do you agree that biomedical waste should be segregated into different categories	72 (97.3)
Do you feel that safe management of biomedical waste is not an issue at all	60 (81.1)
Do you feel that safe management of biomedical waste is the responsibility of the institution and not the individual	57 (77.0)
Do you feel that safe management of biomedical waste is an additional burden of work	64 (86.5)
How often do you recommend cleaning of biomedical waste at your department	15 (20.3)
Do you feel that biomedical waste management should compulsorily be made part of an undergraduate curriculum	66 (89.2)
Do you think your knowledge regarding biomedical waste management is adequate	5 (6.8)
Do you think that you require any further training on biomedical waste management	67 (90.5)

Moreover, approximately 81.1% of participants felt that safe management of BMW was not an issue at all. More than 77.0% felt that BMW management was also the responsibility of an individual, and the majority of them agreed that handling BMW was not an additional burden of work. However, only 20.3% of them recommended daily cleaning of the BMW. Approximately 89.2% agreed that BMW management should be compulsorily in the curriculum for undergraduates. Surprisingly, only 6.8% felt that they had adequate knowledge of BMW management, but the majority of them thought that they needed further training.

Practice on BMW

Table 4 shows that most participants had a high level of practice: 71.4% pathology, 64.1% UPT and haematology, 57.9% microbiology and parasitology and 50.0% emergency.

Table 4. Descriptive results of practice on BMW management

Practice	N (%)
Does your institute have a collaboration with waste management companies	60 (81.1)
Does your department have a collaboration with waste management companies	37 (50.0)
Do you dispose all kinds of waste into general garbage	71 (95.9)
Do you segregate the biomedical waste according to different categories	69 (93.2)
Do you reuse a needle that has been used	71 (95.9)
Do you use first aid kits immediately after contact with biomedical waste that can be contagious	7 (9.5)
Do you report any injuries associated with a sharp waste that was improperly disposed of	50 (67.6)
Where do you dispose of cotton and other items contaminated with blood	74 (100.0)
Where do you dispose of sharp waste	73 (98.6)
Where do you dispose of cotton, gauze and other items contaminated with blood	72 (97.3)
Where do you dispose of pharmaceutical waste	26 (35.1)
How do you dispose of hazardous liquid waste	49 (66.2)

Table 5. Level of knowledge, attitudes and practices among laboratory technologists

Categories	Pathology N (%)	UPT & Haematology N (%)	Microbiology & Parasitology N (%)	Emergency N (%)
Knowledge Level				
High	2 (14.3)	5 (12.8)		
Moderate	11 (78.6)	31 (79.5)	18 (94.7)	2 (100.0)
Low	1 (7.1)	3 (12.8)	1 (5.3)	
Attitude Level				
High	11 (78.6)	33 (84.6)	11 (57.9)	2 (100.0)
Moderate	3 (21.4)	6 (15.4)	7 (36.8)	
Low				
Practice Level				
High	10 (71.4)	25 (64.1)	11 (57.9)	1 (50.0)
Moderate	4 (28.6)	14 (35.9)	8 (42.1)	1 (50.0)
Low				

Table 5 shows the percentage of correctly answered questions on practice. A total of 81.1% of them knew that HUSM collaborated with BMW management companies. Almost all of the participants did not dispose of all clinical waste into general garbage; rather, they segregated BMW according to its categories. Approximately 95.9% of them did not reuse a needle that had been used. Sadly, 9.5% of the participants used a first aid kit immediately after contact with BMW, which can be contagious. Up to 67.6% of participants reported injuries associated with sharp waste. However, the majority of them correctly practiced segregating BMW into different categories, such as cotton and gauze contaminated with blood, pharmaceutical waste, hazardous liquid waste and other waste.

Association between Demographic Data and Knowledge, Attitudes and Practices

Pearson's chi-square test was used to examine the associations between demographic data and knowledge, attitudes and practices. Based on the test, there was no association between demographic data and knowledge and practice. However, Table 6 shows that the demographic data on department categories were found to be associated with attitudes regarding BMW management (p = 0.024).

Correlations between Knowledge, Attitude and Practice Scores

Spearman's correlation test was used to determine the correlations between knowledge and attitude scores, knowledge and practice scores and attitude and practice scores on BMW management. As a result, Table 7 shows that there were no significant differences between the knowledge and attitude scores or between the attitude and practice scores. In contrast, there was a significant, weak and positive correlation between knowledge and practice scores (r = 0.26, p = 0.029).

Table 6. Association between demographic data and attitude

Demographic Variables		Attitude, N (%)			χ^2	*p* value
		High	Moderate	Low		
Department	Pathology	11 (78.6)	3 (21.4)		13.029	0.024*
	UPT & Haematology	33 (84.6)	6 (15.4)			
	Microbiology & Parasitology	11 (57.9)	7 (36.8)	1 (5.3)		
	Emergency		2 (100.0)			

* Fisher's exact test.

Table 7. Correlations between knowledge, attitude and practice scores

Variables	*R*	*p* value
Knowledge with Attitude Score	0.21	0.079
Knowledge with Practice Score	0.26	0.029*
Attitude with Practice Score	0.14	0.225

* Spearman's correlation test.

Table 8. Comparison of knowledge, attitude and practice scores with selected departments

Median (IQR)					
Variable	**Pathology**	**UPT & Haematology**	**Microbiology & Parasitology**	**Chi-square Statistic (df)**	***p*-value**
Knowledge	11.00 (3.25)	11.00 (7.00)	11.00 (4.00)	2.717 (3)	0.437
Attitude	8.50 (5.00)	6.00 (1.00)	8.00 (5.00)	20.486 (3)	<0.001*
Practice	17.50 (5.25)	19.00 (5.00)	19.00 (5.00)	0.994 (3)	0.803

* Kruskal-Wallis test.

Comparison of Knowledge, Attitude and Practice Scores with Selected Departments

The Kruskal-Wallis test was used to compare the knowledge, attitude and practice scores on biomedical waste management between selected departments. As a result, Table 8 shows that the knowledge and practice scores between selected departments were not significantly different. Meanwhile, the attitude scores between selected departments were significantly different ($p < 0.001$). After the post hoc test, the UPT and Haematology department had significantly higher attitude scores (6, IQR 1) than the others.

Discussion

This study showed that the majority of participants had a moderate level of knowledge on BMW management. Most of them were diploma holders, and their knowledge was limited due to a lack of information in the curriculum. Madhukumar and Ramesh (2012) stated that knowledge about BMW rules among healthcare personnel was satisfactory. The majority of them knew that clinical waste was hazardous, and most of them were aware of BMW management and handling rules. This finding was aligned with Mehta et al. (2018), who found that most of them were aware of why BMW was dangerous. Almost all participants were not aware of IMAGE. This was surely due to a lack of training on BMW management. Similarly, in another study by Malini and Eshwar (2015), almost all laboratory technologists had not received any training on BMW management. Surprisingly, the majority did not know the maximum time limit for BMW storage. A previous study also did not find the correct knowledge on the maximum time limit for BMW storage (Mehta et al. 2018).

Regarding attitudes, the majority of them had strong attitudes towards BMW management. In the present study, Ahmed Yar et al. (2018) stated that the overall satisfactory attitude score of nurses was higher than that of laboratory technicians due to their intensive patient care, greater responsibilities and more involvement. Positive findings showed that the majority of them opined that segregated BMW was considered an individual responsibility, and most of them felt that handling BMW was not a burden or work. Malini and Eshwar (2015) found that handling BMW involved teamwork and that everyone was responsible for safe disposal. However, participants felt that they had inadequate knowledge and recommended undergraduate curricula. Therefore, necessary knowledge can be present at the early stage (Njagi et al. 2012).

Most of the participants had a high level of practice towards BMW management. In contrast, a previous study stated that nurses practised BMW management better than laboratory technologists (Shafee et al. 2010). Nurses spent the most time in the clinical ward and handling patients (Sekar et al. 2018). Then, the majority were aware that their institutions collaborated with waste management companies. However, a minority admitted that they disposed of all BMW in the general garbage. The results matched those of this study due to a lack of training, low interest in training programmes and patient overload in hospitals (Hakim et al. 2014). Minor populations do not report injuries caused by improper disposal. A study by Radha (2012) showed that only 13.3% of cases were reported. They were unaware of the formal system of injury reporting that existed in their institution. Almost all participants did not reuse a needle, knew to segregate BMW and knew to treat liquid waste.

An association between department and attitude was found in this study. Different amounts of BMW were generated in different departments. According to Gupta et al. (2009), the attitude towards BMW depends on the amount of waste generated and its contribution to the quantity of waste. A significant weak and positive correlation was found between knowledge and practice scores. Participants had more knowledge, but they had limited practice handling BMW. Similarly, Wai et al. (2005) found a significant association between knowledge and practice with a correlation of 0.39.

There was a significant difference in attitude scores between the selected departments. Post hoc results showed that the UPT and haematology departments had significantly higher attitude scores than other departments because all subjects participated in this study, whereas other departments had subjects who did not agree to participate. The reason for lower KAP values among nurses than laboratory technologists revealed that the majority of the participants were busy with a work schedule and unable to participate (Sekar et al. 2018).

CONCLUSION

This study was conducted to assess the level of knowledge, attitude and practice relating to BMW management among laboratory technologists at selected departments in the Hospital Universiti Sains Malaysia. In conclusion, laboratory technologists in HUSM among selected departments had moderate knowledge and strong attitudes and practices. The results revealed that there was an association between department and attitude on BMW management. Moreover, knowledge and practice scores were significantly positively and weakly correlated. Last, this study revealed that the attitude score between selected departments was significantly different. The UPT and Haematology department had significantly higher attitude scores compared to others.

Recommendation

There is a need for intensive training programmes at regular intervals to train them repeatedly with a special interest on newcomers and regularly familiarise them with updating BMW management to improve their knowledge, attitudes and practices. In addition, orientation programmes for newcomers are required to understand the function of the hospital. Moreover, strict daily monitoring and supervision are recommended in hospitals for waste management activities. Furthermore, any injuries that occur during work must be reported to the person in charge of BMW management or BMW management committee.

Limitations

This study did not assess every aspect of knowledge, attitude and practice questions regarding BMW management. Therefore, future studies should involve an interventionregarding BMW management and risk assessments.

Acknowledgments

We would like to acknowledge the Universiti Sains Malaysia, all the professionals, health personnel and participants who supported this study.

References

Ahmed Yar Mohammed, D. A. B., Muhammad Muqeet, U., Amal Ali, A M. & Fatma, S. A. A. (2018). Knowledge, attitude and practice of biomedical waste management among health care personnel in a secondary care hospital of al buraimi governorate, sultanate of oman.

Global Journal of Health Science, 10(3): 70-82. doi: 10. 5539/gjhs.v10n3p70.

Department of Environment. (2009). *Guidelines on the handling and management of clinical waste in Malaysia*. Retrieved on March 13, 2019 from https://www.doe.gov.my/portalv1/wpcontent/uploads/ 2010/07/anagement_Of_Clinical_Wastes_In_Malaysia__2__0pdf.

Gupta, S., Boojh, R., Mishra, A. & Chandra, H. (2009). Rules and management of biomedical waste at Vivekananda Polyclinic: A case study. *Journal Waste Management*, 29(2): 812-819.

Hakim, S. A., Mohsen, A. & Bakr, I. (2014). Knowledge, attitude and practice of health-care personnel towards waste disposal management at Ain Shams University Hospital, Cairo. *Eastern Mediterranean Journal*, 20: 5.

Madhukumar, S. & Ramesh, G. (2012). Study about awareness and practices about health care waste management among hospital staff in a Medical College Hospital, Bangalore. *International Journal of Basic Medical Science*, 3(1): 1-5.

Malini, A. & Eshwar, B. (2015). Knowledge, attitude and practice of biomedical waste management among health care personnel in tertiary care hospital in Puducherry. *International Journal of Biomedical Research*, 6(3):172-176.

Metha, T. K., Shah, P. D. & Tiwari, K. D. (2018). A Knowledge, Attitude and Practice study of biomedical waste management and bio-safety among healthcare workers in a Tertiary Care Government Hospital in Western India. *National Journal of Community Medicine India*, 9(5): 327.

Njagi, A. N., Oloo, A. M., Kithinji, J. & Kithinji, J. M. (2012). Knowledge, attitude and practice of health-care waste management and associated health risks in the two teaching and referral hospitals in Kenya. *Journal Community Health*, 37: 1172-1177. doi: 10.1007/s 10900-012-9580-x.

Park, K. (2011). *Park's Textbook of Preventive and Social Medicine*. 21st edition, Bhanot Phanot Publisher: Jabalpur: 730.

Patil, A. D. & Shekdar, A. V. (2001). Health-care waste management in India. *Journal of Environmental Management*, 63(2): 211-220.

Radha, R. (2012). Assessment of existing knowledge, attitude and practices regarding biomedical waste management among the health care workers in a tertiary care rural hospital. *International Journal of Health Sciences and Research*, 2(7): 17.

Sachan, R., Patel, M. L. & Nischal, A. (2012). Assessment of the knowledge, attitude and practice s regarding biomedical waste management amongst the medical and paramedical staff in tertiary health care centre. *International Journal of Scientific and Research Publications*, 2(7):1-6.

Sanjeev, R., Kuruvilla, S., Subramaniam, R., Praschant, P. S. & Gopalkrishan, M. (2014). Knowledge, attitude and practices bout biomedical waste management among healthcare personnel in dental colleges in Kothamangalam: a cross-sectional study. *Health Sciences*, 3: 4-7.

Sathaphy, S. & Pandhi, R. K. (1998). *Manual for hospital waste management at AIMS Hospital*, New Delhi.

Sekar, M., Swapna, M. & Easow, J. M. (2018). A study on knowledge, attitude and practice of biomedical waste management among health care workers in a tertiary care hospital in Puducherry. *Original Research Article*, 5(1): 57-60. doi: 10.18231/2394-5478. 2018.0011.

Shafee, M., Kasturwar, N. B. & Nirupama, N. (2010). Study of Knowledge, Attitude and Practices Regarding Biomedical Waste among Paramedical Workers. *Indian Journal of Community Medicine*, 35(2): 369-370.

Wai, S., Tantrakarnapa, K., & Huangprasert, S. (2005). Knowledge, attitude and practice of Myanmar migrant in Maesot District towards the environmental sanitation conditions. *Thai Environmental Engineering Journal,* 19:19.

Yadannavar, M. C., Berad, A. S. & Jagirdar, P. B. (2010). Biomedical waste management: A study of knowledge, attitude and practices in a tertiary health care institution in Bijapur. *Indian Journal of Community Medicine*, 35(170):1.

In: The Influence of Ecosystem Services … ISBN: 978-1-53619-977-2
Editors: Hasmah Abdullah et al.

Chapter 2

KNOWLEDGE, AWARENESS AND PRACTICE OF CLIMATE CHANGE AMONG COMMUNITIES IN KUALA LUMPUR

***Naimah M. F.*[1] *and Haliza A. R.*[2,*]**
[1]Department of Environmental and Occupational Health,
Faculty of Medicine and Health Sciences,
Universiti Putra Malaysia, Selangor, Malaysia
[2]Institute for Social Science Studies, Putra Infoport,
Universiti Putra Malaysia, Selangor, Malaysia

ABSTRACT

Cities are responsible for 75% of global carbon dioxide emissions, making them a major contributor to climate change (CC). Since the urban population is the first to experience the impacts of climate change, their knowledge, attitude, and practise (KAP) were assessed. A self-administered questionnaire has been used to conduct a cross-sectional analysis of 232 respondents aged 15 and above. Five out of eleven districts in Kuala Lumpur (KL) were selected by simple random

[*] Corresponding Author's E-mail: dr.haliza@upm.edu.my.

sampling, while subjects were recruited from community settings using convenient sampling. The relationship between KAP scores was investigated using Spearman correlation tests. To determine t, the Chi-Square test was used. There was an association between knowledge and awareness (r = 0.302), knowledge and practise (r = 0.142) and awareness and practise (r = 0.196). The educational level had a significant association with knowledge and awareness. The community had knowledge about CC, however, they are not aware of what actions to be taken. The government shall disseminate information about CC through mass and social media. Combinations of top-down and bottom-up approaches are important for adaptation and mitigation strategies.

Keywords: knowledge, awareness, practice, climate change, community

INTRODUCTION

Climate change is defined as the change in the state of the climate that can be identified by changes that persist for an extended period, typically decades or longer (Intergovernmental Panel on Climate Change (IPCC) 2014). According to Haliza (2011), climate change is an alteration in the statistical distribution of weather over periods of time that range from decades to millions of years. Matsumoto (2019) emphasized that human socioeconomic activities, particularly the use of fossil fuels, leads to climate change. Evidence of climate change can be seen from the occurrence of extreme events. Heat stress, extreme precipitation, inland and coastal flooding, landslides, air pollution, drought and water scarcity pose risks in urban areas for people, assets, economies and ecosystems (IPCC 2014).

Humans and their unsustainable activities are the most significant contributors to climate change; therefore, there is a need for the community to have knowledge and awareness of climate change and other environmental issues. Hence, several studies were conducted to assess public awareness and perceptions of climate change in various parts of the world. Common knowledge on climate change is very important to alter people's attitudes, which will indirectly be reflected in reducing the adverse effects of climate change (Al Mutairi and Tang 2017).

However, to date, studies assessing knowledge, awareness and practice are still less pronounced in Malaysia; thus, this study intends to fill this knowledge gap. The purpose of this study is to identify the level of knowledge, awareness and practice relating to climate change among urban communities in Kuala Lumpur. This study is very important because the capital city of Kuala Lumpur was described as getting hotter by 0.6°C per decade, which is the world's highest value so far reported for the urban heat island effects through 2005 (Ramakreshnan et al., 2018). The outcome of studies on climate change done in Malaysia is representative to a certain extent of other tropical countries, bearing in mind that climate is a very highly variable phenomenon (Norzaida et al., 2015). The knowledge derived from the research could also be disseminated and shared with communities of similar climate conditions.

METHODOLOGY

A cross-sectional study was conducted in five (5) out of eleven (11) districts of Kuala Lumpur. The five districts included Bukit Bintang, Titiwangsa, Wangsa Maju, Bandar Tun Razak and Cheras. To select the districts, simple random sampling was employed using fishbowl techniques. Five districts were chosen to represent the Kuala Lumpur area. The data collection involved five public places (supermarkets, restaurants, recreational parks, LRT stations and clinics/hospitals) in each of the five districts; thus, ten questionnaires for each place in each district were distributed. This was to ensure that the number of respondents was equally distributed among all selected districts. Community members aged 15 years old and above who had lived in Kuala Lumpur for at least one year were eligible to be study participants (Taghizadeh et al., 2012). The sample size was calculated using the formula of proportion for one group (Lemeshow et al., 1990). Hence, the total sample size was 236 respondents altogether; however, only 232 respondents fully completed the survey.

Ethical approval for the study was obtained from the Ethics Committee Universiti Putra Malaysia (JKEUPM-2018-363).

For the research instrument, a self-administered questionnaire was used. The questionnaire was adapted and modified from Yale University's "Six Americas" survey and the World Health Organization (WHO) Revised Baseline KAP Report. The questionnaire consists of 5 sections: Section A (Sociodemographic Characteristics), Section B (Knowledge on Climate Change), Section C (Awareness of Climate Change), Section D (Practice on Climate Change) and Section E (Recommendations). To ensure the reliability and consistency of the questionnaire, a pretest was conducted, and the overall Cronbach's alpha of the items in the questionnaire was reported to be 0.74, which shows that the questionnaire has good reliability.

The data were analysed using SPSS version 22. To determine if the data were normally distributed, the Kolmogorov-Smirnov test was used. Then, descriptive tests were employed to analyse the sociodemographic information, knowledge, awareness and practice levels among the respondents. In addition, descriptive tests were used to determine the sources of information about climate change as well as the factors that prevent communities from taking action on climate change. The Spearman correlation test was used to identify the associations between scores of knowledge, awareness and practice relating to climate change. Next, to examine the associations between educational level and knowledge and awareness level, chi-square tests were used.

RESULTS AND DISCUSSION

Sociodemographic Characteristics

Of the respondents, 53% were female and 47% were male. In terms of age, most of the respondents were within the age of 15-24 years old (35.8%) and between 25 and 40 years old (44.0%). In terms of occupation, the majority (40.9%) of participants worked in the private sector.

Regarding education level, 33.6% of the respondents had a low level of education (received no education, primary school and secondary school), whereas the other 66.4% had a high level of education (diploma or matriculations, bachelor's degree, master's and PhD). Many of the respondents (35.3%) had lived in Kuala Lumpur all their lives.

Knowledge, Awareness and Practice Level Relating to Climate Change

With regard to knowledge level, the majority (84.9%) of participants had moderate knowledge. The findings of this study are similar to those of a study performed among communities in China, where the general public showed a moderate level of knowledge and understanding of climate change (Al Mutairi and Tang 2017). However, the results of this study are different from those of a study conducted in Pakistan, in which only half of the urban community (55.3%) had knowledge about climate change (Ashraf et al., 2016).

Most respondents had very high (60.3%) and high (34.5%) levels of awareness, whereas none of the respondents reported low awareness. The results are presented in Table 1. Similar observations have been reported in a study done in India, where the general urban population was aware of climate change and human activities as significant causes of it (Harshal et al., 2011). Similarly, Al Buloshi and Ramadan (2015) reported that awareness among the public in Oman was fairly high.

The majority (75.0%) of participants had moderate practice. A study assessing climate change awareness among communities in the Philippines showed similar findings to the present study, where the respondents' practice of mitigating actions of climate change was moderate (Lesley and Lalevie 2019). According to Arbaat et al., (2013), the respondents expressed that they were not aware of what actions to take to deal with climate change issues. This explains why their practice was still at a moderate level.

Table 1. Level of knowledge, awareness and practice relating to climate change

Variables	Number of respondents (n = 232)	Percentage (%)
Knowledge on CC		
Low	23	9.9
Moderate	197	84.9
High	12	5.2
Awareness on CC		
Low	0	0
Moderate	12	5.2
High	80	34.5
Very High	140	60.3
Practice on CC		
Poor	28	12.1
Moderate	174	75.0
Good	30	12.9

Associations between Knowledge, Awareness and Practice Relating to Climate Change

There was a positive and weak association between knowledge and awareness of climate change ($r = 0.302$, $p = 0.028$). A positive and weak association was found between knowledge and practice ($r = 0.142$, $p = 0.03$). The association between awareness and practice was positive and weak ($r = 0.196$, $p = 0.003$).

There was a significant association between knowledge and awareness of climate change. This result was similar to previous research done in Ghana, Africa, where there was a positive and weak correlation between knowledge and awareness of climate change (Awusi and Asare 2016). Aminrad et al., (2013) stated that people who are knowledgeable about the environment and its associated issues will become more aware of the environment.

There was also a significant relationship between knowledge and practice level, and this result was similar to a study done by Arbaat et al., (2013), where there was a significant relationship between environmental knowledge and practices. The state of one's knowledge about an issue is associated with their practices (Aminrad et al., 2013). Even though an association was found between knowledge and practice, the correlation was weak, and this finding was similar to the study of Jamilah et al., (2015) in which a weak relationship was found between students' level of knowledge and sustainable environmental practices. In addition, Arbaat et al., (2013) stated that there was a significant correlation between awareness and practices towards the environment, whereas our present study also had similar results. The level of awareness was translated into the tendency to engage in sustainable practices.

Our finding was supported by a study done in Bangladesh in which people with higher educational levels had more knowledge about climate change (Kabir et al., 2016). According to Diaz-Quijano et al., (2018), education level is the key determinant of knowledge. The respondents with a higher intellectual level might reflect their interest in environmental knowledge, especially when it is associated with their lives.

Therefore, our study suggests that environmental education should be strengthened in the school curriculum to increase knowledge about climate change and its adaptation measures. Education is an essential element of the community response to climate change. It helps people understand climate change causes, motivates them to change unsustainable practices towards environmentally friendly behaviour and helps them adapt to impacts.

The results also showed that educational level was a significant factor related to awareness level, and this finding was similar to those of Lee et al., (2015) and Harshal et al., (2011). Aminrad et al., (2011) stated that educational level influenced awareness of the environment.

Association between the KAP Levels and Sociodemographic Characteristics

In terms of knowledge level, there were no significant differences found across age group, gender, race, family income, marital status, occupation and years of living. However, there was a significant association between knowledge and educational level ($p = 0.028$). For the association between educational level and knowledge, the Pearson chi-square statistics was 7.15, and the degrees of freedom was 2. A total of 16.7% of the expected cell frequencies had a value less than 5. Most of the respondents with low and high education levels had a moderate level of knowledge, as most of the respondents reported that there was less information available.

In regard to awareness, there were no significant differences found across age group, gender, race, family income, marital status and occupation. However, there were significant differences found across educational levels ($p = 0.004$). The Pearson chi-square statistic was 10.827, and the degrees of freedom was 2. However, 16.7% of the expected cell frequencies had a value less than 5. The respondents with a high education level had very high awareness of climate change, $n = 104$ (67.5%). For the low education group, most of them had high awareness, $n = 38$ (48.7%).

Sources of Information

The most common media in which the community members acquired information about climate change were news (82.3%), television (78%) and the Internet (72%). The least reported sources of information were the workplace (24.6%), government (28%) and books/magazines (30.6%).

Per the survey conducted among the community in Belize, Central America, the study also reported that 66.5% of the respondents perceived television as the most effective channel to obtain information about climate change. A similar finding was reported in a previous study by Bojovic (2014) assessing climate change awareness among citizens of Macedonia, where 67% of the respondents gained information through television and 71% preferred the Internet the most. Meanwhile, in a study conducted in Indonesia, Sulistyawati et al., (2018) found that more than half of the respondents (53.54%) claimed that family was their favourite source of information, whereas only 15.16% chose the Internet to obtain information. These differences were due to accessibility to technologies and differences in lifestyle between regions. For some parts of the country, the people still hold on to family values, regardless of the advancement in the digital era. However, for some other countries, people depend on the Internet and technologies to find information. In Malaysia, access to the Internet is easy, and the Internet price rate offered by telecommunication companies is affordable. According to Malaysian Communications and Multimedia Commission (2018), Internet users in Malaysia increased from 76.9% in 2016 to 87.4% in 2018. Conventional media, such as television and the news, were still favoured by most community members across regions because of their high degree of trustworthiness and credibility. This reflected the reason why the respondents of this study preferred news, television and the Internet as platforms to gain information.

Therefore, in this study, it is suggested that the government focus on giving information to the public about climate change through conventional sources, such as news, television and radio. In addition, the Internet is another powerful source of information where awareness campaigns or nonformal education could be spread through Facebook, Twitter, YouTube and other social media sites.

Factors that Prevent Communities from Taking Action on Climate Change

Most respondents (61.6%) claimed that they were not aware of what actions to take to reduce climate change. A total of 46.6% reported that they did not receive enough information about climate change. In addition, 19.4% of respondents felt that climate change was not a concern of the community. Next, 5.6% did not perceive mitigating climate change as their responsibility.

This result was supported by a study from Nepal that produced similar outcomes. Despite youth awareness of climate change, they were still uncertain about the adaptation and mitigation strategies implemented by the government (Devkota and Phuyal 2017). According to a report by the Japan-Caribbean Climate Change Project (2016), the KAP survey found that community members had not taken action because there was not enough information, and 40.4% said that they were not aware of what to do about the problems. Six percent stated that climate change should not be the community's concern, and the rest felt that it was not their responsibility.

Even though there were still some respondents in this study who perceived that climate change issues were not supposed to be the community's concern, and some thought that it was not their responsibility, the percentages were still relatively smaller. In general, the community still considered the environmental issue, but due to insufficient information given, ignorance subsisted. This showed that the government should sensitize the public to climate change issues and disseminate more information related to climate change via all possible channels.

CONCLUSION

From this study, it can be concluded that the community had very high awareness of climate change issues. However, their knowledge about the problems was at a moderate level, as was their knowledge of the practice level. This shows that even though the community was aware that climate change was happening, they had no idea what actions to take due to a lack of information received. Malaysia has made significant progress in establishing a legal framework for the implementation of climate change mitigation. However, improving education, training and public awareness on climate change is also an important measure. Therefore, efforts to mitigate climate change impacts must be made with the combination of both top-down and bottom-up approaches to ensure that any policies or programmes on climate change that are implemented by the government will be successful. The government, through its relevant agencies, may conduct public awareness campaigns and community education. Any provisions or efforts being implemented by the government should include community participation because they are the largest driving forces that could make small efforts large and effective when they are done collectively.

ACKNOWLEDGMENTS

The author would like to thank the community of Kuala Lumpur who willingly participated and, therefore, made this study possible and successful.

REFERENCES

Al Buloshi, A. S. and Ramadan, E. (2015). Climate change awareness and perception amongst the inhabitants of Muscat Governorate, Oman. *American Journal of Climate Change*, 4 (4): 330- 336.

Al Mutairi, K. and Tang, H. (2017). Public knowledge and awareness of climate changes among people in China. *Current World Environment*, 12(2): 231-236.

Aminrad, Z., Syed Zakaria, S. Z. and Hadi, A. S. (2011). Influence of age and level of education on environmental awareness and attitude: Case study on Iranian students in Malaysian Universities. *Social Sciences*, 6(1):15-19.

Aminrad, Z., Syed Zakaria, S. Z., Hadi, A. S. and Sakari, M. (2013). Relationship between awareness, knowledge and attitudes towards environmental education among secondary school students in Malaysia. *World Applied Sciences Journal*, 22(9): 1326-1333.

Arbaat, H., Norshariani, A. R. and Sharifah, I. S. S. A. (2013). The level of environmental knowledge, awareness, attitudes and practices among UKM students. *Journal of Environment & Behavior*, 33(5): 687-707.

Ashraf, M. Q., Wahab, P. and Adeel, M. (2016). Knowledge, attitude and perception about climate change among people of urban area in Attock, Pakistan. *International Journal of Agricultural and Environmental Research*, 2(4):333-338.

Awusi, E. and Asare, K. (2016). Climate change knowledge and awareness creation in relation to the media among senior high students in Birim Central Municipal, Ghana. *Journal of Environmental Science, Toxicology and Food Technology*, 10(9): 83-89.

Devkota, N., and Phuyal, R. K. (2017). An analysis of Nepalese youth understanding level on climate change. *Asian Journal of Economic Modelling*, 5(8): 342-353.

Diaz-Quijano, F. A., Martinez-Vega, R. A., Rodriguez-Morales, A. J., Rojas-Calero, R. A., Luna-González, M. L. & Díaz-Quijano, R. G. (2018). Association between the level of education and knowledge, attitudes and practices regarding dengue in the Caribbean region of Colombia. *BMC Public Health*, 18: 1-10.

Haliza, A. R. (2011). Climate change phenomena: Is human in danger? *Health and the Environment Journal*, 2(1): 54-58.

Harshal, T. P., Chawla, P. S., Fernandez, K., Singru, S. A., Khismatrao, D. and Pawar, S. (2011). Assessment of awareness regarding climate change in an urban community. *Indian Journal Occupational and Environmental Medicine*, 15(3):109-112.

Intergovernmental Panel on Climate Change. (2014). *Climate Change 2014 Fifth Assessment Report.* Retrieved on 2 March 2020 from https://www.ipcc.ch/site/assets/uploads/2018/05/SYR_AR5_FINAL_full_wcover.pdf.

Jamilah, A., Shuhaida, M. N. and Nurzali, I. (2015). Investigating students' environmental knowledge, attitude, practice and communication. *Asian Social Science*, 11: 1911-2025.

Kabir, M. I., Rahman, M. B., Smith, W., Lusha, M. A. F., Azim, S. and Milton, A. B. (2016). Knowledge and perception about climate change and human health: findings from a baseline survey among vulnerable communities in Bangladesh. *BMC Public Health*, 16: 261-266.

Lee, T. M., Markowitz, E. M., Howe, P. D., Ko, C. Y. and Leiserowitz, A. A. (2015). Predictors of public climate change awareness and risk perception around the world. *Nature Climate Change*, 5:1014-1020.

Lesley, C. L. and Lalevie, C. L. (2019). Knowledge, attitudes, practices, and action on climate change and environmental awareness of the twenty-two villages along the River Banks in Cagayan de Oro City, Philippines: PART II. *Acta Scientific Agriculture*, 3:114-125.

Malaysian Communications and Multimedia Commission. (2018). *Internet users survey 2018*. Retrieved on March 2, 2020 from https://www.mcmc.gov.my/skmmgovmy/media/General/pdf/Internet-Users-Survey-2018.pdf.

Matsumoto, K. (2019). Climate change impacts on socioeconomic activities through labor productivity changes considering interactions between socioeconomic and climate systems. *Journal of Cleaner Production*, 216: 528-541.

Norzaida, A., Zalina, M. D., Norazizi M. and Syafrina, A. H. (2017). Climate change impact on coastal communities in Malaysia. *Journal of Advanced Research Design*, 33(1): 1-7.

Ramakreshnan, L., Aghamohammadi, N., Fong, C. S., Ghaffarianhoseini, A., Ghaffarianhoseini, A., Li, P. W., Norhaslina, H. and Nik Meriam, S. (2018). A critical review of urban heat island phenomenon in the context of Greater Kuala Lumpur, Malaysia. *Sustainable Cities and Society*, 39: 99-113.

Sulistyawati, S., Mulasari, S. A. and Sukesi, T. W. (2018). Assessment of knowledge regarding climate change and health among adolescents in Yogyakarta, Indonesia. *Journal of Environmental and Public Health*, 9: 1-7.

Taghizadeh, A. O., Navidi, I., Hosseini. M., and Ammari, H. (2012). Knowledge, attitude and practice of Tehran's inhabitants for an earthquake and related determinants. *PLoS Currents Disasters*, Aug 6. Edition 1. doi: https://10.1371/4fbbbe1668eef.

In: The Influence of Ecosystem Services ... ISBN: 978-1-53619-977-2
Editors: Hasmah Abdullah et al.

Chapter 3

KNOWLEDGE, ATTITUDE AND PRACTISE OF *ZIKA* DISEASE AMONG ENVIRONMENTAL AND OCCUPATIONAL HEALTH STUDENTS AT PUBLIC UNIVERSITIES IN MALAYSIA

Nur Shaherah A. R.[1,2] ***and Haliza A. R***[1,2,*]

[1]Department of Environmental and Occupational Health, Faculty of Medicine and Health Sciences, Universiti Putra Malaysia, Serdang, Selangor, Malaysia

[2]Institute for Social Science Studies, Putra Infoport, Universiti Putra Malaysia, Serdang, Selangor, Malaysia

ABSTRACT

The Zika virus is a vector-borne disease that shares the same transmission mode as dengue fever. It has the potential to cause neurological problems in children. Several cases of Zika disease had been recorded in Malaysia before December 2016. Then, in order to respond to epidemics and pandemics, it was necessary to determine awareness,

[*] Corresponding Author's E-mail: dr.haliza@upm.edu.my.

attitude, and practise. Our aim was to investigate Zika disease awareness, attitudes, and practises among students of Environmental and Occupational Health at a Malaysian public university. A cross-sectional survey and self-administered questionnaire was distributed using universal sampling. Purposive sampling was then used to choose the stage of study year. The knowledge, attitude and practise on Zika disease were measured using adapted survey tools by World Health Organization. Descriptive analysis showed the level of knowledge, attitude and practise of 172 students Environmental and Occupational Health were at a moderate level. However, the result of an association between socio-demographic and practise, showed a significant difference ($p = 0.005$) between age and practise towards prevention of Zika disease. Meanwhile, a correlation test was conducted to identify if there was any correlation between knowledge, attitude and practise. Then, there was a weak significant positive correlation between knowledge and attitude ($r = 0.205$, $p = 0.007$) and between practise and attitude ($r = 0.549$, $p = 0.001$). The result also had shown no significant difference in knowledge, attitude and practise between first and fourth-year students. By determining the level of knowledge, attitude and practise of Zika disease among university students, it can provide baseline data and guides to health authorities in order to plan preventive measure programs in future.

Keywords: *Zika* disease, knowledge, attitude, and practise, public university, Malaysia

INTRODUCTION

Zika disease (ZIKD) is an infectious disease caused by the *Zika* virus (ZIKV). The transmission of ZIKV is the same as that of dengue virus (DV), which is transmitted primarily via the bite of *Aedes* mosquitoes, *Aedes aegypti* (Carneiro and Travassos 2016). Currently, *Aedes* mosquitoes are very persistent across Asia and America because daily temperature fluctuations affect the transmission of *Aedes aegypti* (Lambrechts et al. 2011). This vector can disseminate easily in tropical and subtropical areas (Kang et al. 2010). Obviously, the higher concentration of the vector in particular areas can also cause rapid dissemination of ZIKV. Moreover, climate change can also cause the transmission of vector-borne diseases. By 2100, the transmission of vector-borne diseases in new areas is estimated to increase by 1.0 – 3.5°C. However, climate change varies across countries, and the health risks exposed by the population are also different (Githeko et al. 2000). *Aedes aegypti* is one of the vectors that can transmit most vector-borne diseases (chikugunya, dengue, *Zika*) and is very sensitive to temperature changes. When the temperature rises, the larvae take a shorter time to mature. Thus, this situation can make the larvae produce more offspring during the transmission period (Rueda et al. 1990).

In addition to transmission via mosquito-borne virus, blood transfusion and perinatal transmission can also cause ZIKD (Besnard et al. 2014). Moreover, the probable non-vector transmission of ZIKV is most likely caused by sexual transmission (Foy et al. 2011). Research will be conducted to identify how long ZIKV can stay in the semen and vaginal fluids of people infected by ZIKV (CDC 2016). ZIKV was known globally when outbreaks occurred in Brazil and other countries in May 2015. These outbreaks have gained worldwide attention because of the increased incidence of microcephaly and Guillain-Barré syndrome (GBS). The World Health Organization (WHO) also strongly suspects that there is a relationship between ZIKD and microcephaly and GBS. Therefore, the WHO declared ZIKV a "Public Health Emergency of International Concern" on February 1, 2016. This can help in monitoring and managing ZIKD cases in each country that has a higher possibility of having outbreaks of ZIKD (WHO 2016).

Since many cases of ZIKD related to microcephaly and GBS have been reported, it can impact the country itself, especially its economy (Mlakar et al. 2016). Based on the International Vaccine Access Centre (2017), the government of Brazil continues to make investments in preventive and control measures of *Aedes aegypti*. The government will also continue to make debt repayments because of the absence of significant financial inflows to finance outbreak control and surveillance infrastructure. In addition, countries will lose productivity because children with microcephaly may not have the ability to achieve their full cognitive potential. This is because children who are generally born with microcephaly will have a disability in cognitive and physical development.

Therefore, these children are less likely to attend school, resulting in lower educational achievement (Constenia et al. 2017).

Next, there were 33 countries involved in the outbreak between January 2014 and February 2016. Brazil and Colombia are noticeable countries that have been reported to have the highest numbers of ZIKD cases (Petersen et al. 2016). However, the transmission of ZIKV in Southeast Asia has become an emerging health problem. This is because vectors that carry ZIKV, which is an arboviral disease, are very common (Wiwanitkit 2016). Several cases of ZIKD have been reported in the Philippines (Alera et al. 2015), Cambodia (Heang et al. 2012) and Indonesia (Kwong et al. 2013). In Singapore, the total cases of ZIKD surpassed 400 through October 5, 2016 (Hui 2016). This total number will increase the level of worry among Malaysians. Malaysia is located near the equatorial line like Singapore. Additionally, many Malaysians who live in Johor Province tend to work in Singapore, and they will be exposed to ZIKV frequently.

Currently, 8 cases of ZIKD have been recorded in Malaysia since 2016 (Abdullah 2016 a). According to the Ministry of Health (MOH), in 2016, the first case of *Zika* infection was detected in Malaysia in a 47-year-old, who was infected with ZIKV in Singapore. Then, the following case was in Sabah. This case shows that there was an *Asian lineage* similar to the genome in French Polynesia (Abdullah 2016b). The other 3 cases occurred because they worked in an infected area, such as Singapore. Next, Malaysia confirmed an eighth Zika virus case involving a 67-year-old man who first showed symptoms in late November. Therefore, the MOH issued a directive as part of the "Updated *Zika* Alert" to manage and monitor the transmission of ZIKV in Malaysia (MOH 2016). This is important to take into account before an outbreak occurs because it can cause a great deal of damage to the health, environment or property of the country. Research preparedness of ZIKD is required not only in epidemiology and laboratory diagnosis but also in awareness and increases in the knowledge of the population. According to the WHO (2016), regarding outbreaks of disease and health emergencies, people's perceptions and beliefs can determine how they take advice on practise (Yimer et al. 2014). Thus, when populations have knowledge about ZIKD, they can practise the prevention step of ZIKD and directly decrease the number of ZIKD cases.

This article, therefore, serves to address this gap in the literature by identifying the level of knowledge, attitude and practise (KAP) relating to ZIKD; determining the associations between sociodemographic variables, knowledge and attitude with practise towards prevention; determining the correlations between KAP; and comparing the KAP between first- and fourth-year students of environmental and occupational health (ENVOCH).

METHODOLOGY

Study Participants

This research was carried out as part of a cross-sectional study to estimate the KAP on ZIKD among students of ENVOCH in public universities in Malaysia. Universal sampling was used to obtain a list of public universities that offered an ENVOCH course in Malaysia. There were two universities that offered a linear course of ENVOCH: Universiti Putra Malaysia (UPM) and Universiti Sains Malaysia (USM). Then, purposive sampling was used to select first- and fourth-year ENVOCH students at these public universities in Malaysia. The total population involved in this research was 172 students.

Survey Questionnaire

The questionnaire was adapted from survey tools provided by the WHO, 'Knowledge, attitude, and practise surveys *Zika* disease and potential complications'. The respondents were assessed using self-administration, which consisted of 5 sections. Section A consists of sociodemographic data and general information about ZIKD, such as when the participants first heard about ZIKD and where they heard about it. Sections B, C and D consist of knowledge, attitudes and practises towards

ZIKD. Last, section E includes recommendation questions about ZIKD. This section was used to identify the effective way to prevent ZIKD and determine the most appropriate medium to disseminate information about ZIKD.

The knowledge question was a multiple-choice question. Each correct response was given a score of 1, and incorrect responses were given a score of 0. Therefore, the maximum total score was n = 37, and the minimum total score was n = 0. The modified Bloom cut-off point was 80 to 100% for good knowledge, 60 to 79% for moderate knowledge and <60% for poor knowledge (Fabrigar et al. 2006). Thus, the score of the respondents' knowledge can be categorized as high (n = 30 to 37), moderate (n = 22 to 29), and poor (n = 0 to 21). The attitudes and practises of the respondents were assessed using Likert scales. There were 10 questions for the attitudes and practises section. An attitude score was given based on their answer, such as strongly agree 4, agree 3, not sure 2, disagree 1, and strongly disagree 0. Meanwhile, a practise score was given as always 4, often 3, sometimes 2, seldom 1, and never 0. However, the score was reversed when assessing the negative attitudes and practises questions. The modified Bloom's cut-off point was also used in the attitudes and practises section, and it was categorized as high (n = 32 to 40), moderate (n = 24 to 31), and low (n = 0 to 23).

Statistical Analysis

The data were analysed using the Statistical for Social Sciences (SPSS) version 21. Different statistical tests were used based on different objectives. Descriptive analysis was used to determine the frequency (percentage) of participants falling into high, moderate or low levels of KAP. The chi-square test was used to determine if there was an association between sociodemographic variables, knowledge and attitudes with practises towards the prevention of ZIKD. In addition, Spearman's rho correlation was used to identify the correlation between KAP among ENVOCH students. To compare the knowledge and attitudes of first- and fourth-year ENVOCH students, an independent sample t-test was used. Meanwhile, the Mann-Whitney U test was used to compare the practises of students between the first and fourth years.

RESULTS AND DISCUSSION

The results here were discussed based on the questionnaire section as well as the sequence of objectives. A total of 172 students responded to the self-administered questionnaire.

Regarding the general question about ZIKD, approximately 55.2% of the ENVOCH students first heard about ZIKD in the last few months. A total of 42.4% of the respondents first heard about ZIKD many years ago. The respondents mostly first heard about ZIKD many years ago and in the last few months because the recent outbreak occurred in 2016. However, 2.3% first heard about ZIKD just recently. Figure 1 shows when the students first heard about ZIKD.

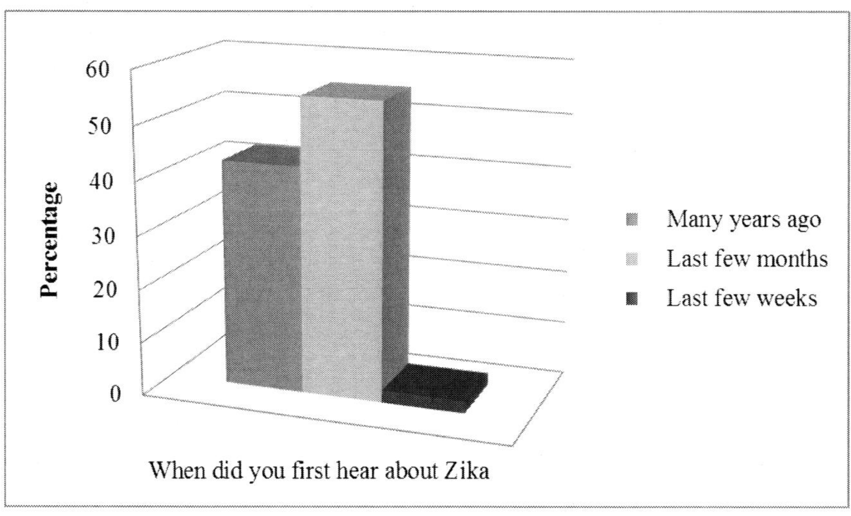

Figure 1. When students first heard about ZIKD.

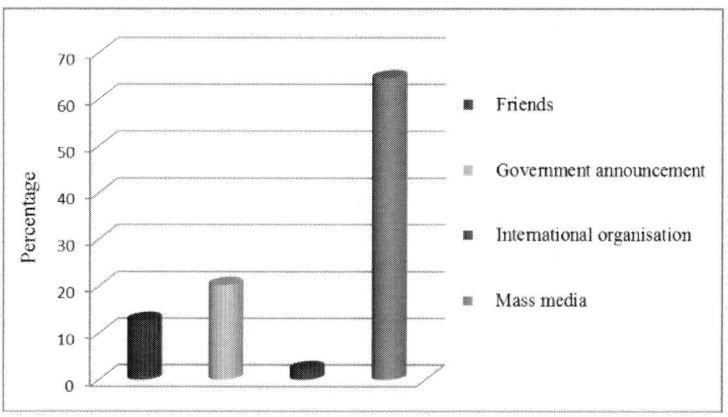

Figure 2. Where the students first heard about ZIKD.

Meanwhile, as shown in Figure 2, students first heard about ZIKD from mass media (64.5%), followed by government announcements (20.3%), friends (12.8%), and international organizations (2.3%).

Table 1 shows the level of KAP relating to ZIKD among ENVOCH students.

Table 1. Level of KAP relating to ZIKD among ENVOCH students

Variable	Frequency (N)	Percentage (%)
Knowledge		
High	51	29.7
Moderate	**84**	**48.8**
Low	37	21.5
Total	172	100.0
Attitude		
High	48	27.9
Moderate	**85**	**49.4**
Low	39	22.7
Total	172	100.0
Practise		
High	6	3.5
Moderate	**94**	**54.7**
Low	72	41.9
Total	172	100.0

Referring to the table above, 84 (48.8%) of the students had a moderate level of knowledge. Then, 51 (29.7%) students had a high level of knowledge on ZIKD, and 37 (21.5%) had only a low level of knowledge. The knowledge of the students about ZIKD was still moderate because the public attention was only gained in 2016 in Brazil. Therefore, the students needed time to understand the knowledge about *Zika*.

Regarding attitudes, 85 (49.4%) of the ENVOCH students had a moderate level of attitude. Meanwhile, 48 (27.9%) of the respondents had a high level of attitude, and 39 (22.7%) had a low level of attitude. Some studies have shown that the attitude of the person depends on their knowledge (Abdullah 2016a). Thus, this study also shows that attitudinal trends depend on the knowledge of the students.

Next, this study shows that only 6 (3.5%) students had a high level of practise in the prevention of ZIKD, and 72 (41.9%) of them have low levels of practise. Even though they were university students, their practise in the prevention of ZIKD was still low. This is because they were not interested in the prevention measures of ZIKD and the impact of the *Zika* outbreak in their community (Al-Zurfi et al. 2015).

One of the reasons for the low practise among university students was the time constraints to perform preventive measures. This study is consistent with the findings of Shuaib et al. (2010), who reported a high level of knowledge but a low level of practise (Shuaib et al. 2010). Meanwhile, the study by Acharya et al. (2005) shows contrasting findings of a low level of knowledge but a high level of practise (Acharya et al. 2005). However, this study showed that 94 (54.7%) students had a moderate level of practise towards the prevention of ZIKD.

Table 2 below shows the association between sociodemographic variables and practise level. Age showed a significant association with practise in the prevention of ZIKD ($p = 0.005$). However, gender, previous education level and race did not have a significant association with practise because the p-value was more than 0.05. Females showed a high level of practise, 34 (19.8%), compared to males, 14 (8.1%). These findings supported study by Regitz-Zagrosek (2012), which showed that women seek health care more frequently than men (Regitz-Zagrosek 2012). This is because the behaviour of the person differs based on their gender (Ngun et al. 2012). In line with a study in Malaysia, a previous study of ZIKV conducted among healthcare students in medicine, pharmacies and dentistry showed that female students had more knowledge of the progress of ZIKV than male students (Ahmad et al. 2017). Respondents under 20 years old had a lower level of practise, 26 (15.1%), than respondents between 21 and 30 years old, 13 (7.6%). Patients under 20 years old were considered adolescents. Meanwhile, those between 21 and 30 years old were considered adults (WHO 2013). Thus, these differences can influence practise and act differently. Adults usually take an action by considering the effect and consequences in the future related to their life events. However, the adolescents tend to take any action before considering a long-term perspective (Karaman 2007). Previous education level from matriculation/pre-university had a high level of practise, 39 (22.7%), relating to the prevention and control programme than education from a diploma, 7 (4.1%), and STPM/STAM, which was 2 (1.2%).

Next, students with the race of Malay had a high level of practise, 43 (25.0%), compared to Chinese 3 (1.7%), Indian 1 (0.6%), and other races 1 (0.6%). This is because most of the respondents were Malay. Thus, race cannot be a factor that affects ZIKD prevention programmes. A study by Ahmad et al. (2017), on the other hand, found that Indians had more adequate knowledge than other races.

The association between knowledge and practise among first-year and fourth-year students was determined for both universities, UPM and USM. Table 3 shows that the p-value was 0.928 ($p > 0.05$). Hence, the null hypothesis was accepted because there was no association between knowledge and practise among first-year ENVOCH students. Meanwhile, the p-value in Table 13 was 0.213. Thus, there was no association between knowledge and practise among fourth-year ENVOCH students. This shows a linear result for both years because neither has an association between knowledge and practise.

This study's findings were in contrast with those of Al-Adhroey et al. (2010) and Dike et al. (2006), who reported that there was a positive association between knowledge and practise or behaviour (Dike et al. 2006; Al-Adhroey et al. 2010).

Table 2. The association between sociodemographic variables and practise level

Demographic variable	Practise, n (%)			Total	$^aX^2$	*p-value
	High	Moderate	Low			
Gender					0.512	0.774
Male	14 (8.1)	20 (11.6)	10 (5.8)	44 (25.6)		
Female	34 (19.8)	65 (37.8)	29 (16.9)	128 (74.4)		
Total	48 (27.9)	85 (49.4)	39 (22.7)	172 (100)		
Age					14.774	**0.005**
<20	13 (7.6)	37 (21.5)	26 (15.1)	76 (44.2)		
21 – 30	35 (20.3)	47 (27.3)	13 (7.6)	95 (55.2)		
>30	0 (0.0)	1 (0.6)	0 (0.0)	1 (0.6)		
Total	48 (27.9)	85 (49.4)	39 (22.7)	172 (100)		
Previous education					7.463	0.113
STPM/STAM	2 (1.2)	1 (0.6)	0 (0.0)	3 (1.7)		
Matriculation/ Pre-university	39 (22.7)	80 (46.5)	37 (21.5)	156 (90.7)		
Diploma	7 (4.1)	4 (2.3)	2 (1.2)	13 (7.6)		
Total	48 (27.9)	85 (49.4)	39 (22.7)	172 (100)		
Race					7.388	0.286
Malay	43 (25.0)	81 (47.1)	37 (21.5)	161 (93.6)		
Chinese	3 (1.7)	1 (0.6)	0 (0.0)	4 (2.3)		
Indian	1 (0.6)	2 (1.2)	0 (0.0)	3 (1.7)		
Others	1 (0.6)	1 (0.6)	0 (0.0)	4 (2.3)		
Total	48 (27.9)	85 (49.4)	39 (22.7)	172 (100)		

[a]Pearson chi-square

*p-value < 0.05 is significant

Table 3. Association between knowledge and practise on ZIKD among first-year ENVOCH students

Variable	Practise, n (%)			Total	${}^{a}X^2$	*p-value
Knowledge, n (%)	High	Moderate	Low			
High	5 (5.5)	13 (14.3)	6 (6.6)	24 (26.4)	1.872	0.928
Moderate	7 (7.7)	22 (24.2)	15 (16.5)	44 (48.4)		
Low	5 (5.5)	11 (12.1)	7 (7.7)	23 (25.3)		
Total	17 (18.7)	46 (50.5)	28 (30.8)	91 (100.0)		

[a]Pearson chi-square
*p-value < 0.05 is significant

Table 4. Association between knowledge and practise on ZIKD among fourth-year ENVOCH students

Variable	Practise, n (%)			Total	${}^{a}X^2$	*p-value
Knowledge, n (%)	High	Moderate	Low			
High	12 (14.8)	14 (17.3)	1 (1.2)	27 (33.3)	5.823	0.213
Moderate	16 (19.8)	18 (22.2)	6 (7.4)	40 (49.4)		
Low	3 (3.7)	7 (8.6)	4 (4.9)	14 (17.3)		
Total	31 (38.3)	39 (48.1)	11 (13.6)	81 (100.0)		

[a]Pearson chi-square
*p-value < 0.05 is significant

From Table 5 below, the p-value of the association between attitude and practise on ZIKD among first-year ENVOCH students was 0.033, where the value was less than 0.05. Therefore, there was a significant association between attitude and practise among first-year ENVOCH students. Even though they were still first-year students, their attitude towards practise on preventive measures was good. One of the factors that can contribute to developing the attitude of a person is additional knowledge and learning from family or parents (UNICEF 2014).

Table 6 shows that there was no association between attitude and practise on ZIKD among fourth-year ENVOCH students because the *p*-value was 0.083 and more than 0.05. Thus, the null hypothesis was accepted.

Table 5. Association between attitude and practise on ZIKD among first-year ENVOCH students

Variable	**Practise, n (%)**			**Total**	**$^aX^2$**	***p-value**
Attitude, n (%)	**High**	**Moderate**	**Low**			
High	2 (2.2)	0 (0.0)	0 (0.0)	2 (2.2)	10.481	0.033
Moderate	4 (4.4)	21 (23.1)	12 (13.2)	37 (40.7)		
Low	11 (12.1)	25 (27.5)	16 (17.6)	52 (57.1)		
Total	17 (18.7)	46 (50.5)	28 (30.8)	91 (100.0)		

[a] Pearson chi-square

*p-value < 0.05 is significant

Table 6. Association between attitude and practise on ZIKD among fourth-year ENVOCH students

Variable	**Practise, n (%)**			**Total**	**$^aX^2$**	***p-value**
Attitude, n (%)	**High**	**Moderate**	**Low**			
High	0 (0.0)	2 (2.5)	1 (1.2)	3 (3.7)	8.255	0.083
Moderate	16 (19.8)	17 (21.0)	9 (11.1)	42 (51.9)		
Low	15 (18.5)	20 (24.7)	1 (1.2)	36 (44.4)		
Total	31 (38.3)	39 (48.1)	11 (13.6)	81 (100.0)		

[a] Pearson chi-square

*p-value < 0.05 is significant

The correlation between KAP among ENVOCH students was determined using Spearman's correlation because the data were not normally distributed. As seen in Table 16, there was a weak significant positive correlation between knowledge and practise (r = 0.205, p = 0.007) and between practise and attitude (r = 0.549, p = 0.000). The study by Wan

Rozita et al. (2006) also showed the same correlation results (Wan Rozita et al.2006). However, there was no significant result between knowledge and practise (r = 0.030, p = 0.692) because the p-value was more than 0.01. This study was different from De Pretto et al. (2015), whose findings show the correlation between knowledge, attitude and practise (De Pretto et al. 2015). In addition, the respondents assumed that they faced time constraints to practise the prevention measure. Moreover, Heera and Parajuli (2016) also reported similar findings. The subject took time and commitment to practise the prevention and control programme consistently (Heera & Parajuli 2016).

Table 7. Correlation between KAP among ENVOCH students

Variable	**Coefficient correlation, r (p-value)**		
	Knowledge	**Attitude**	**Practise**
Knowledge	-	0.205 (0.007)*	0.030 (0.692)
Attitude	0.205 (0.007)*	-	0.549 (0.000)*
Practise	0.030 (0.692)	0.549 (0.000)*	-

*Spearman's rho correlation
*p-value < 0.01 is significant

To compare the level of KAP between first- and fourth-year ENVOCH students, all KAP scores were subjected to the Shapiro-Wilk test to perform a normality test. All the scores for knowledge and attitude were assumed not normal because the p-values were less than the 5% level of significance. Meanwhile, the score for practise was assumed to be normal because the value was more than the 5% level of significance.

Based on Table 8, there was no significant difference in KAP between first-year and fourth-year ENVOCH students. These findings were consistent with Arbiol et al. (2016), which showed no association between knowledge and attitude scores relating to infectious disease (Arbiol et al. 2016). However, this was in contrast with the study by Bakhoum et al. (2014), in which the findings showed that there was a significant difference in knowledge between people with senior degrees and junior degrees at the university (Bakhoum et al. 2014). The study conducted at the Faculty of Medicine in Malaysia also shows that fourth-year students have more knowledge than third-year students. One of the factors that contributed to this finding was the knowledge received by first- and fourth-year ENVOCH students at the same time because the outbreaks of Zika in Brazil captured public attention in 2016. This situation also causes attitudes and practises to develop at the same time. Thus, there was no significant difference between the two groups.

In Section E of the questionnaire, recommendation questions about preventive measures of ZIKD were asked. Figure 3 shows the percentage of the most effective way selected by students to prevent ZIKD in Malaysia.

Table 8. Comparison between first- and fourth-year ENVOCH students

Variable	Mean (SD)/Median (IQR)		$^{a}Z/^{b}t$	*p-value
	First Year	Fourth Year		
Knowledge	23.00 (24.0)	23.00 (23.0)	- 0.972[a]	0.331
Attitude	24.00 (23.0)	24.00 (23.0)	- 1.011[a]	0.312
Practise	23.82 (6.26)	25.17 (6.3)	- 1.406[b]	0.162

[a]Mann-Whitney U Test
[b]Independent Samples t-test
**p*-value < 0.05 is significant

Figure 3 shows that only 5% agreed to disseminate information about ZIKD on printed media/mass media, and 1% agreed to increase the relationship with related organizations. Then, 35% agreed to avoid visiting countries affected by ZIKV. However, 59% agreed that the most effective way to prevent ZIKD in Malaysia was by enhancing knowledge and attitudes about ZIKD and knowing the right preventive measure. According to Sharma et al. (2000), health education to the community, such as students at the university, and proper training of health personnel can strengthen surveillance and go a long way with the control of emerging disease (Sharma et al. (2000). Therefore, increasing their knowledge can help them to improve their attitudes and practises towards the prevention of ZIKD. The involvement of communities, such as university students, in the prevention of infectious disease can help reduce the burden of disease in their area (Tambo et al. 2014).

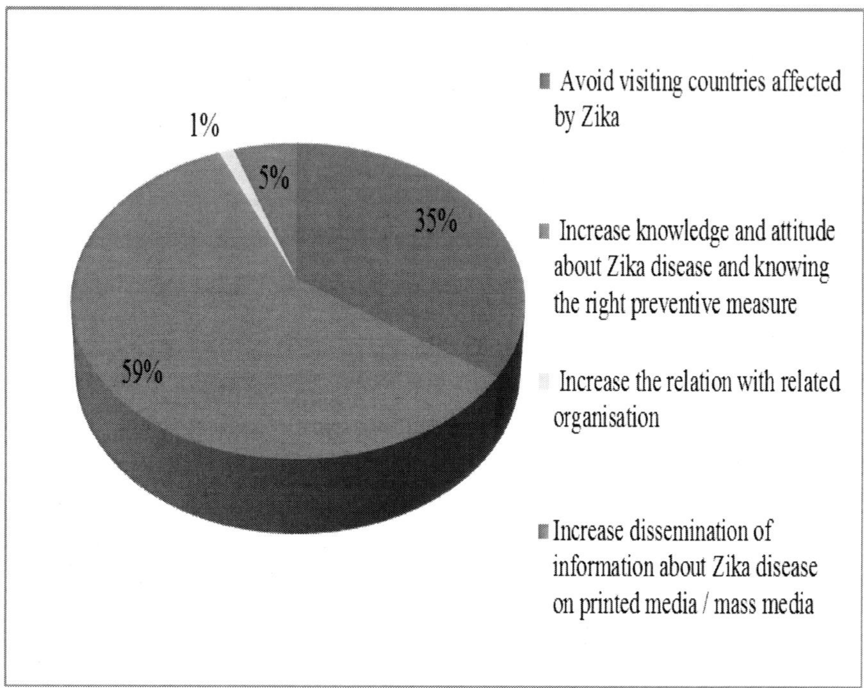

Figure 3. The most effective way to prevent ZIKD in Malaysia.

Table 9. Most appropriate medium for disseminating information about ZIKD

Variable	**n**	%
Mass media	60	34.9
Printed media	4	2.3
Social media	108	62.8
Total	172	100.0

The results of the descriptive analysis of the most appropriate medium for disseminating information about ZIKD that students chose are tabulated in Table 9. There were 108 (62.8%) students who chose social media, 60 (34.9%) who chose mass media and 4 (2.3%) who chose printed media as the medium for disseminating information.

Students currently prefer social media as a medium to disseminate information because it is convenient for everyone to use. Additionally, almost every person in this decade has a smart phone and social media account. Thus, they can easily access their accounts using their smart phones. The study conducted by Duggan et al. (2016) showed that 82% of students in college have an active account of *Facebook* (Duggan et al. 2016). A study on social media conducted by Osterrieder (2013) showed that people can keep up to date with current research and broader issues and popular science. Most importantly, it is highly personalized by individuals (Osterrieder 2013). For example, the social media that are used now are *Facebook*, *Twitter*, Google+, *YouTube*, and *Instagram*. These platforms can open opportunities for others to engage in conversation or dialogue and are not limited only to scientists (Gallice and Bora 2015).

Conclusion

This study found that the level of KAP of the ENVOCH students was still at a moderate level. However, there was a significant difference between age and practise regarding prevention measures of ZIKD. In addition, this study also discovered a significant correlation between knowledge and attitude and between attitude and practise. Since ZIKD spread globally, attention and awareness from the public, including university students, is needed. This is because the epidemic of ZIKV in the future is still unpredictable. It may have the potential to spread worldwide like dengue fever because it comes from the same vector-borne transmission. Therefore, if university students still ignore ZIKD and do not take any preventive steps at an early stage, it can cause an increase in ZIKD cases in the future. Thus, it was important to perform the research preparedness, especially in knowing KAP. Determining the level of KAP can guide health authorities, such as governments, policy makers or non-government organizations (NGOs), to plan the appropriate vector-borne control programme and activity, especially among university students in Malaysia

Recommendations

The recommendation given by this research was based on the needs of students in a university setting in Malaysia.

Use Social Media Effectively to Disseminate Information about ZIKD

Currently, most people, especially younger people, tend to have their own social media accounts, such as *Facebook*, *Twitter*, *blogs*, and *vlogs*. Thus, it was important to use social media platforms to disseminate information, such as control and prevention steps of ZIKD, symptoms of ZIKD and facts about ZIKD. The information can be delivered in an interesting way if health authorities can make a video about the issue and health problem related to vector-borne infection. By using social media frequently, it can help to close the gap in vector-related knowledge, especially ZIKD.

Continue the Campaign of Health Promotion in the University

Health promotion should be conducted continuously in university areas and not just during epidemics or outbreaks. Continuous campaigns can increase awareness and help to change practise. Because knowledge alone is insufficient to promote the prevention of vector-borne diseases like ZIKD, the behaviour cannot become a habit without continuous training.

The students at the university can benefit from this campaign by sharing their knowledge and experience during health promotion to the non-student population, such as the community in residential areas or schools, because they always engage in activities with that population.

Amend the Existing Subject of Epidemiology at the University

The subject epidemiology should focus on current infectious diseases, such as *Zika*. The learning outcome should include mode of transmission, symptoms of the affected person, treatment and prevention and control measures. Thus, this information will guide and help students to increase their knowledge and develop good attitudes and practises.

References

Abdullah, N. H. (2016). Press Statement KPK 18 Disember 2016 – Current situation of Zika in Malaysia: 8th Case, *Situasi terkini Zika di Malaysia: Kes Zika ke-8*. Retrieved on May 5, 2020 from the desk of the Director-General of Health Malaysia website: https://kpkesihatan.com/2016/12/18/kenyataan-akhbar-kpk-18-disember-2016-situasi-terkini-zika-di-malaysia-kes-zika-ke-8/.

Abdullah, N. H. (2016). Press Statement KPK 28 Ogos 2016 – *Current situation of Zika*, *Situasi terkini virus Zika*. Retrieved on May 5, 2020 from the desk of the Director-General of Health Malaysia website: https://kpkesihatan.com /2016/08/28/kenyataan-akhbar-kpk-28-ogos-situasi-terkini-virus-zika-di-malaysia/.

Acharya, A., Goswami, K., Srinath, S. & Goswami, A. (2005). Awareness about dengue syndrome and related preventive practises amongst residents of an urban resettlement colony of south Delhi. *Journal of Vector Borne Diseases*, *42*(3), 122–127.

Ahmad, A., Bhagavathulha, A. S. & Jamshed, S. Q. (2017). A Cross-sectional survey on the knowledge and attitudes towards Zika Virus and its prevention among residents of Selangor, Malaysia. *Journal of Pharmacy Practise and Community Medicine*, *3*(2), 81–89.

Al-Adhroey, A. H., Nor, Z. M., Al-Mekhlafi, H. M. & Mahmud, R. (2010). Opportunities and obstacles to the elimination of malaria from Peninsular Malaysia: knowledge, attitudes and practises on malaria among aboriginal and rural communities. *Malaria Journal*, *9*, 137. doi:https://doi.org/10.1186 /1475-2875-9-137.

Alera, M. T., Hermann, L., Tac-An, I. A., Klungthong, C., Rutvisuttinunt, W., Manasatienkij, W., Villa, D., Thaisomboonsuk, B., Velasco, J. M., Chinnawirotpisan, P. & Lago, C. B. (2015). Zika virus infection, Philippines, 2012. *Emerging Infectious Diseases*, *21*(4), 722–724. doi:https://doi.org /10.3201/ eid2104.141707.

Al-Zurfi, B. M., Fuad, M. D., Abdelqader, M. A., Baobaid, M. F., Elnajeh, M., Ghazi, H. F., Ibrahim, M. H. & Abdullah, M. R. (2015). Knowledge, attitude and practise of dengue fever and health education programme among students of Alam Shah Science School, Cheras, Malaysia. *Malaysian Journal of Public Health Medicine*, *15*(2), 69–74.doi:https://doi.org /10.12691/ajphr-1-2-2.

Arbiol, J., Orencio, P. M., Romena, N., Nomura, H., Takahashi, Y. & Yabe, M. (2016). Knowledge, attitude and practises towards leptospirosis among Lakeshore Communities of Calamba and Los Baños, Laguna, Philippines. *Agriculture*, *6*(18), 1–12. doi:https://doi.org/10.3390/agriculture6020018.

Bakhoum, A. Y., Bachmann, M. O., El Kharrat, E. & Talaat, R. (2014). Assessment of knowledge, attitude, and practise of risky sexual behavior leading to HIV and sexually transmitted infections among Egyptian substance abusers: A cross-sectional study. *Advances in Public Health*, 1–8. doi: https://doi.org/10.1155/2014/701861.

Besnard, M., Lastère, S., Teissier, A., Cao-Lormeau, V. M. & Musso, D. (2014). *Evidence of perinatal transmission of zika virus*, French Polynesia, December 2013 and February 2014. *Eurosurveillance*, *19*(13), 8–11. doi:https://doi.org /10.2807/1560-7917.ES2014. 19.13.20751.

Carneiro, L. A. M. & Travassos, L. H. (2016). Autophagy and viral diseases transmitted by *Aedes aegypti* and *Aedes albopictus*. *Microbes and Infection*, *18*(3), 169–171. doi: https://doi.org/10.1016/ j.micinf.2015.12.006.

Communicable Disease Control. (2016). *Transmission & risks through mosquito bites from mother to child through sex through blood transfusion through laboratory and healthcare setting exposure*. Retrieved from CDC website: https://www.cdc.gov/zika/transmission/

Constenia, D., Broucker, G. de. & Campo, J. M. del. (2017). *The potential economic impact of the zika virus*. IVAC Blog, The potential economic impact of the zika virus mosquitos breed in standing water, pp. 1–2.

De Pretto, L., Acreman, S., Ashfold, M. J., Mohankumar, S. K. & Campos-Arceiz, A. (2015). The link between knowledge, attitudes and practises in relation to atmospheric haze pollution in Peninsular Malaysia. *PLoS ONE*, *10*(12), 1–18. doi:https://doi.org/10.1371/journal.pone.0143655.

Dike, N., Onwujekwe, O., Ojukwu, J., Ikeme, A., Uzochukwu, B. & Shu, E. (2006). Influence of education and knowledge on perceptions and practises to control malaria in Southeast Nigeria. *Social Science and Medicine*, *63*(1), 103–106. doi:https://doi.org/10.1016/ j.socscimed. 2005. 11.061.

Duggan, M., Page, D. & Manager, S. C. (2016). *Social Media Update 2016. In Pew Research Center*. Retrieved on May 4, 2020 from http://assets.pewresearch.org/wp-content/uploads/sites/14/2016/11 /10132827/PI_2016.11.11_Social-Media-Update_FINAL.pdf.

Fabrigar, L. R., Petty, R. E., Smith, S. M. & Crites, Jr. S. L. (2006). Understanding knowledge effects on attitude-behavior consistency: the role of relevance, complexity, and amount of knowledge. *Journal of Personality and Social Psychology*, *90*(4), 556–577. doi: https://doi.org/10. 1037/0022-3514.90.4.556.

Foy, B. D., Kobylinski, K. C., Foy, J. L. C., Blitvich, B. J., da Rosa, A. T., Haddow, A. D., Lanciotti, R. S. & Tesh, R. B. (2011). Probable non–vector-borne transmission of Zika virus, Colorado, USA. *Emerging Infectious Diseases*, *17*(5), 880.

Gallice, G. & Bova, J. (2015). *Social Media: giving science communication a facelift how social media is revolutionizing the way scientists promote their research.* (November 2014): 1–31. doi:https://doi.org/10.7287/peerj. preprints.16.

Githeko, A. K., Lindsay, S. W., Confalonieri, U. E. & Patz, J. A. (2000). Climate change and vector-borne diseases: a regional analysis. *Bulletin of the World Health Organization*, *78*(9), 1136–1147.

Heang, V., Yasuda, C. Y., Sovann, L., Haddow, A. D., da Rosa, A. P. T., Tesh, R. B. & Kasper, M. R. (2012). Zika virus infection, Cambodia, 2010. *Emerging Infectious Diseases*, *18*(2), 349–351. doi:https:// doi.org /10.3201/eid1802. 111224.

Heera, K. & Parajuli, S. (2016). Dengue awareness and practise among the people living in haraincha village development committee of Eastern Nepal. *Birat Journal of Health Sciences*, *1*(1), 38-46. https://doi.org/ 10.3126/bjhs. v1i1.17095.

Hui, M. C. (2016). Only one Zika cluster left; total number of cases surpasses. The Straits Times, pp. 1–2.

Kang, P., Liao, M., Wester, M. R., Leeder, J. S. & Pearce, R. E. (2010). NIH Public Access. *Ratio*, *36*(3), 490–499. doi:https://doi.org /10.1124/dmd.107.016501 .CYP3A4-Mediated.

Karaman, N. G. (2007). Adoção de comportamentos de risco por adolescentes : uma comparação das opiniões de adolescentes e adultos Adopción de comportamientos de riesgo por adolescentes : una comparac. *Paidéia*, *17*(38), 357–364. [Adoption of risk behaviors by adolescents: a comparison of the opinions of adolescents and adults. Adoption of risk behaviors by adolescents: a comparison.]

Kwong, J. C., Druce, J. D. & Leder, K. (2013). Case report: Zika virus infection acquired during brief travel to Indonesia. *American Journal of Tropical Medicine and Hygiene*, *89*(3), 516–517.doi:https:// doi.org/10.4269 /ajtmh.13-0029.

Lambrechts, L., Paaijmans, K. P., Fansiri, T., Carrington, L. B., Kramer, L. D., Thomas, M. B. & Scott, T. W. (2011). Impact of daily temperature fluctuations on dengue virus transmission by Aedes aegypti. *Proceedings of the National Academy of Sciences of the United States of America*, *108*(18), 1–6.doi:https://doi.org/10.1073/ pnas.110137

7108//DC Supplemental.www.pnas.org/cgi/doi/10.1073/ pnas.110137 7108.

Malaysia confirms its 8th Zika virus case. (2016). *Eyewitness News.*

Ministry of Health (MOH). (2016). *Updated zika alert and abdministrative order for compliance and management of zika virus infection*, Updated zika alert" dan arahan pentadbiran untuk pemantauan dan pengurusan jangkitan virus zika.

Mlakar, J., Korva, M., Tul, N., Popović, M., Poljšak-Prijatelj, M., Mraz, J., Kolenc, M., Resman Rus, K., Vesnaver Vipotnik, T., Fabjan Vodušek, V. & Vizjak, A. (2016). Zika virus associated with microcephaly. *New England Journal of Medicine*, *374*(10), 951-958. doi: https://www.nejm.org/doi/full/ 10.1056/ NEJMoa1600651.

Ngun, T. C., Negar, G., Francisco, S. J., Sven, B. & Eric, V. (2012). The genetics of sex differences in brain and behavior. *NIH Public Access*, *32*(2), 227–246. doi:https://doi.org/10.1016/j.yfrne.2010.10.001.

Osterrieder, A. (2013). The value and use of social media as communication tool in the plant sciences. *Plant Methods*, *9*(1), 26. doi:https://doi.org/ 10.1186/ 1746-4811-9-26.

Petersen, E., Wilson, M. E., Touch, S., McCloskey, B., Mwaba, P., Bates, M., Dar, O., Mattes, F., Kidd, M., Ippolito, G. & Azhar, E. I. (2016). Rapid spread of zika virus in the Americas - Implications for public health preparedness for mass gatherings at the 2016 Brazil Olympic Games. *International Journal of Infectious Diseases*, *44*, 11–15. doi:https://doi.org /10.1016/j.ijid.2016.02.001.

Regitz-Zagrosek, V. (2012). Sex and gender differences in health. *EMBO Reports*, *13*(7), 596–603. doi:https://doi.org/10.1038/embor.2012.87.

Rueda, L. M., Patel, K. J., Axtell, R. C. & Stinner, R. E. (1990). Temperature-dependent development and survival rates of *Culex quinquefasciatus* and *Aedes aegypti* (Diptera: Culicidae). *Journal of Medical Entomology*, *27*(5), 892–898.

Sharma, S. N., Raina, V. K. & Kumar, A. (2000). Dengue/DHF: an emerging disease in India. *The Journal of Communicable Diseases*, *32*(3), 175–179.

Shuaib, F., Todd, D., Campbell-Stennett, D., Ehiri, J. & Jolly, P. E. (2010). Knowledge, attitudes and practises regarding dengue infection in Westmoreland, Jamaica. *The West Indian Medical Journal*, *59*(2), 139–146. doi:https://doi.org/10.1055/s-0029-1237430.Imprinting.

Tambo, E., Ugwu, E. C. & Ngogang, J. Y. (2014). Need of surveillance response systems to combat Ebola outbreaks and other emerging infectious diseases in African countries. *Infectious Diseases of Poverty*, *3*(1), 29. doi: https://doi.org /10.1186/2049-9957-3-29.

The United Nations Children's Fund, (UNICEF). (2014). *Study of parental knowledge, attitudes and practises related to early childhood development.*

Wan Rozita, W. M., Yap, B. W., Veronica, S., Muhammad, A. K., Lim, K. H. & Sumarni, M. G. (2006). Knowledge, attitude and practise (KAP) survey on dengue fever in urban Malay residential area in Kuala Lumpur. *Malaysian Journal of Public Health Medicine*, *6*(2), 62–67.

Wiwanitkit, V. (2016). The current status of Zika virus in Southeast Asia. *Epidemiology and Health*, *38*, 1–5. doi:https://doi.org/10.4178/epih .e2016026.

World Health Organization (WHO). (2013). *HIV / AIDS definition of key terms.* Retrieved from World Health Organization website: http://www.who.int/hiv /pub/guidelines/arv2013/intro/keyterms/en/.

World Health Organization. (2016). *Zika virus situation report.* Retrieved from http://www.who.int/emergencies/zika-virus/situation-report/5-february-2016/en/.

Yimer, M., Abera, B., Mulu, W. & Bezabih, B. (2014). Knowledge, attitude and practises of high risk populations on louse- borne relapsing fever in Bahir Dar City, North-West Ethiopia. *Science Journal of Public Health*, *2*(1), 15. doi: https://doi.org/ 10.11648/j.sjph.20140201.13.

In: The Influence of Ecosystem Services … ISBN: 978-1-53619-977-2
Editors: Hasmah Abdullah et al.

Chapter 4

CHANGES IN PARTICULATE MATTER CONCENTRATIONS DUE TO VEHICULAR EMISSIONS AND THEIR INFLUENCE ON URBAN AIR QUALITY IN KELANTAN

***Nur Khairina Syahirah S., Nurulilyana S.,*[*]**
Raja Nur Shafieza R. M. and Nur Fatihah R.
School of Health Sciences, Health Campus,
Universiti Sains Malaysia, Kubang Kerian, Kelantan, Malaysia

ABSTRACT

Roundabouts are designed to replace traffic signalised intersections because they are thought to improve traffic flow, minimise fuel consumption, and reduce noise levels. It is, however, vulnerable to a variety of air pollutants, and one of those is particulate matter. The objective of this research is to determine the changes of selected particulate matter ($PM_{2.5}$ and PM_{10}) concentrations caused by vehicular emissions and their impacts on urban air quality in Kelantan, as well as to compare $PM_{2.5}$ and PM_{10} concentrations between different roundabouts

[*] Corresponding Author's E-mail: nurulilyana@usm.my.

and to determine the relationship between $PM_{2.5}$ and PM_{10} concentrations and vehicle count. $PM_{2.5}$ and PM_{10} concentrations were measured at three selected roundabouts in Kelantan by using DustTrak DRX Desktop 8533. The equipment was located at the centre of the roundabout for two hours per session. Besides, a vehicle counter was used to count vehicles manually according to their type. Data collection was done for seven days continuously at each roundabout. The highest readings were recorded at the roundabout at Pasir Pekan, (336 ± 69.66 μg/m^3) for $PM_{2.5}$ and (352 ± 73.41 μg/m^3) for PM10. There was no significant difference of PM2.5 ($p = 0.345$) and PM_{10} ($p = 0.238$) concentrations between three selected roundabouts. However, there was a significant correlation between the number of vehicles and particulate matter concentrations, which are $PM_{2.5}$ ($p = 0.003$) and PM_{10} ($p = 0.004$). Roundabout at Pasir Pekan is considered the most polluted, as it has the highest mean and median of $PM_{2.5}$ and PM_{10} concentrations compared to the other two roundabouts.

Keywords: air quality, $PM_{2.5}$, PM_{10}, vehicular emissions

INTRODUCTION

Air pollution is a worldwide problem and a major threat to the surrounding environment and human health, especially in low- and middle-income countries. It has been recorded that the eastern Mediterranean, Southeast Asian and western Pacific regions experience much higher exposures than any other regions. The World Health Organization (WHO) has stated that the majority of continents have higher annual $PM_{2.5}$ concentrations in urban areas than in rural areas, except in African and eastern Mediterranean regions (WHO 2018a). Air pollution may be described as contamination of indoor or outdoor air that modifies its natural characteristics and is often not visible to the naked eye (WHO 2018b).

The major sources of air pollutants in Malaysia are mobile sources and stationary and transboundary emissions. Emissions from motor vehicles are one of the examples of mobile sources. Stationary sources include power plants, industrial waste incinerators, and the emission of dust from urban construction work and quarries. Transboundary pollution refers to transported air pollution from forest fires from neighbouring countries. The worst transboundary haze episodes occurred in 1997. Widespread forest fires in Kalimantan and Sumatra were triggered by extremely dry weather due to the El Niño phenomenon. The Klang Valley and Kuching in Sarawak were among the worst hit areas in Malaysia (Dominick et al. 2012). Traffic emissions are recognized as the major cause of air pollution in urban areas and large cities, in addition to domestic, commercial and industrial activities (Gulia et al. 2015; Kanabkaew et al. 2013). The dramatic increase in vehicles in these areas is due to urbanisation. It occurs as people prefer to live in cities to pursue opportunities in various population densities in large cities (Kanabkaew et al. 2013). Badami (2005) and Wang et al. (2010) have stated that vehicular emissions contribute approximately 70-80% of air pollution in megacities in developing countries due to the large number of older vehicles with poor vehicle maintenance, low fuel quality, poor road infrastructure and inadequate inspection and maintenance programs (Wang et al. 2010; Badami 2005).

Emissions from vehicles contain a wide variety of air pollutants, mainly carbon monoxide (CO), carbon dioxide (CO_2), oxides of nitrogen (NO_x), particulate matter (PM) 10 μm (microns) in diameter (PM_{10}), and hydrocarbons (HC) or volatile organic compounds (VOCs). These air pollutants may have a major long-term impact on air quality (Gastaldi et al. 2014; Mandavilli et al. 2008). Particulate matter (PM) was the focus of this study, as the majority of previous studies have focused more on other air pollutants, such as CO and SO_2. PM can be defined as a mixture of tiny solid particles and liquid droplets in the form of organic and inorganic substances found in the air. Generally, particulate matter is divided into two groups based on size. The first group includes PM_{10} with diameters generally between 2.5 microns and 10 microns. Examples of PM_{10} include dust, dirt, soot and smoke. The second group includes $PM_{2.5}$ with diameters of generally 2.5 microns or less that usually form in the atmosphere from the interaction of primary air pollutants (USEPA 2018). Components of PM commonly consist of sulphate, nitrates, ammonia, sodium chloride, black carbon, mineral dust and water (WHO 2018c). There are two types of PM emissions: primary and secondary PM. For primary particulate matter, the particles are emitted directly into the atmosphere in their unmodified form, whereas for secondary particulate matter, the pollutants are partially transformed into particles with the help of photochemical reactions in the atmosphere (EEA 2010). Primary PM is generated from road transportation, combustion (mainly coal burning), and other industrial processes; whereas secondary PM is generated through chemical reactions among different primary particulates in the atmosphere, such as sulphates and nitrates (Wu et al. 2018). The main objective of this study is to determine the changes in particulate matter ($PM_{2.5}$ and PM_{10}) concentrations due to traffic emissions and their influence on urban air quality in Kelantan, Malaysia.

METHODOLOGY

Location of Study

This study was conducted at selected roundabouts in Kelantan, Malaysia. Three roundabouts were chosen for this study, namely, a roundabout in Pasir Pekan, a roundabout in Wakaf Bharu and the Tuan Padang roundabout clock tower in Kota Bharu, as shown in Table 1.

Table 1. Characteristics of selected roundabouts in Kelantan, Malaysia

Roundabout	**Characteristics**			
	Code	Coordinates	Area of the district (km^2)	Population of the district*
Roundabout at Pasir Pekan	R1	6°7’9” N 102°12’57” E	Tumpat - 169.5	147,179
Roundabout at Wakaf Bharu	R2	6°7'23.6"N 102°12'12.5"E		
Tuan Padang roundabout clock tower (Kota Bharu)	R3	6°7'31"N 102°14'15"E	Kota Bharu - 115.6	468,438

Source: Google Maps, 2018.

* Based on ‘The population and housing census of Malaysia in 2010.’

These roundabouts were chosen because they were among the busiest roundabouts in Kelantan, Malaysia. Based on observations prior to data collection, the roads around the roundabout were always congested with vehicles during peak hours. Even police officers were needed to control the traffic, especially in the morning, when people rushed to workplaces or schools. Another contributing factor was the large populations that live in the districts (The population and housing census of Malaysia 2010). Figures 1 to 3 show the locations of each roundabout (Google maps 2018).

Data Collection

Reconnaissance was performed before the actual data collection. The purpose of the reconnaissance was to determine the most favourable placement of DustTrak instrumentation at the roundabouts. Hence, DustTrak for PM measurement was placed in the middle of each roundabout for one hour and then at one of the intersections of the roundabout for another hour. Since both methods produced similar readings, it was then decided to place DustTrak instrumentation in the middle of each roundabout.

Source: Google maps, 2018.

Figure 1. Roundabout at Pasir Pekan (R1).

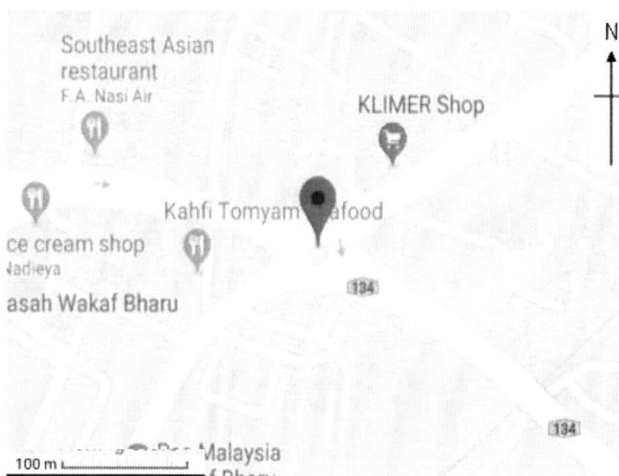

Source: Google maps, 2018.

Figure 2. Roundabout at Wakaf Bharu (R2).

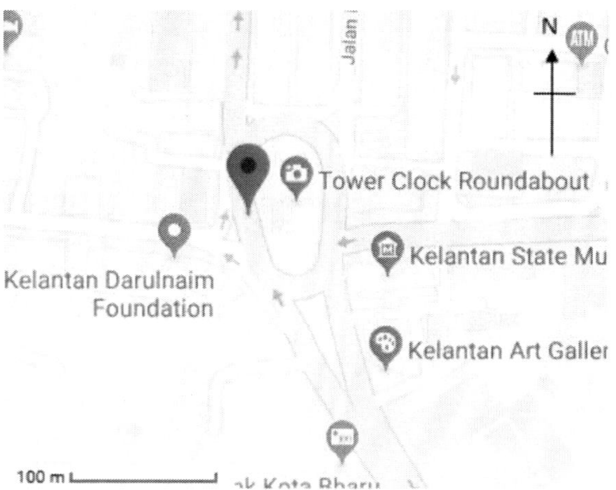

Source: Google maps, 2018.

Figure 3. Tuan Padang roundabout clock tower (Kota Bharu) (R3).

Measurements of $PM_{2.5}$ and PM_{10} were performed at three selected sampling locations by using DustTrak DRX Desktop 8533. The measurements were taken during three peak hours, in which each measurement involved two hours of monitoring.

The three peak hours were as follows: (i) a morning session occurred from 7.00 am to 9.00 am, (ii) an afternoon session occurred from 12.00 noon to 2.00 pm, and (iii) an evening session occurred from 4.00 pm to 6.00 pm. The PM concentrations were monitored every minute during the time interval. All monitoring was performed for 21 days (seven continuous days per sampling site) at each roundabout.

Furthermore, together with these activities, the number of vehicles that passed through the study area (each selected roundabout) was calculated at each intersection to avoid double counting. Figure 4 shows an illustration of sampling activities that involved counting the number of vehicles. The position of the observer at each intersection of the roundabouts is shown. Based on the traffic flow, the number of vehicles was calculated only in one direction (in which the vehicles entered each intersection of the roundabout). Vehicle counting was divided into six types of vehicles, namely, cars, motorcycles, buses, lorries, vans and others, by using vehicle counters.

Statistical Analysis

The data were processed and analysed by using Statistical Package for Social Sciences software (SPSS) version 24. Descriptive statistics were applied to obtain the number of vehicles during peak hours, mean concentration, and minimum and maximum readings of PM concentration for all roundabouts.

A normality test was used to determine if the data collected had a normal distribution. The Kruskal-Wallis test was used to compare $PM_{2.5}$ and PM_{10} concentrations between roundabouts. Spearman's correlation was used to determine the relationship between the number of vehicles and $PM_{2.5}$ and PM_{10} concentrations.

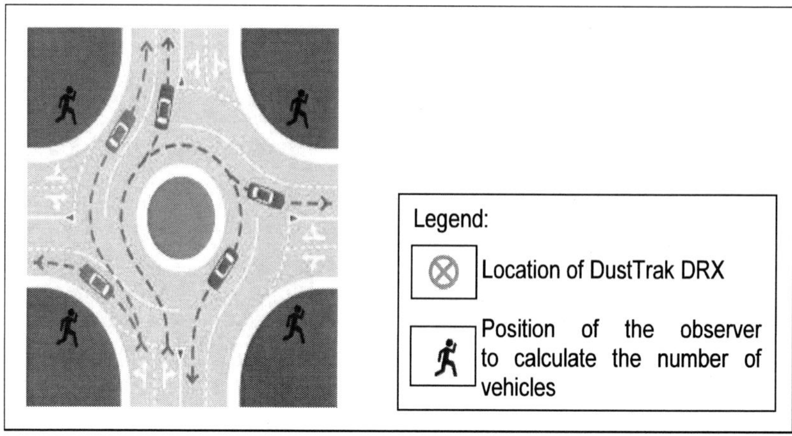

Figure 4. Traffic flow at the roundabout and the position of the observer to calculate the number of vehicles.

Table 2. Number of vehicles for all three roundabouts

Type of Vehicles	Number of Vehicles								
	Minimum			Maximum			Mean		
	R1	R2	R3	R1	R2	R3	R1	R2	R3
Cars	6433	3288	4158	10231	9529	11577	8209.19	5813.43	8589.95
Motorcycles	1844	1655	2055	5763	10374	5094	3688.62	3001.86	3588.00
Buses	21	5	70	54	22	131	30.05	13.33	85.48
Lorries	80	50	42	559	252	252	311.48	128.24	125.62
Vans	90	51	49	263	205	215	192.48	106.48	143.57
Others	4	1	1	48	20	301	16.95	7.19	27.86

* R1 is the roundabout at Pasir Pekan, R2 is the roundabout at Wakaf Bharu, and R3 is the Tuan Padang roundabout at Kota Bharu.

RESULTS AND DISCUSSION

Descriptive Statistics

Number of Vehicles during Peak Hours

A summary of the minimum, maximum and mean data for the number of vehicles is shown in Table 2. The highest mean number of vehicles (n =

8589.95) was cars at R3, and the lowest mean number (n = 7.19) was other vehicles at R2. Other vehicles consisted of vehicles that could not be placed into one of the conventional categories of vehicles (cars, motorcycles, lorries, buses, and vans) and usually had the lowest frequencies. Examples of other types of vehicles included bicycles and trishaws.

Out of all six types of vehicles, as expected, cars had the highest mean for all three roundabouts, since they are the most common form of transportation (refer to Table 2). However, these results contradict those of Srimuruganandam and Shiva Nagendra (2010), because they found that two-wheelers had the highest number of vehicles during both weekdays and weekends. Cars placed next after the two-wheelers (Srimuruganandam and Shiva Nagendra 2010). Moreover, the maximum number of other vehicles (301) was due to a bicycle event at roundabout Kota Bharu on Saturday, 16th Feb 2019.

However, the mean of all vehicles at the Wakaf Bharu roundabout was the lowest and had a large gap compared with the other two roundabouts, except for lorries (refer to Table 3). There was a Chinese New Year Festival, at that time and it was a public holiday. Thus, many people were on holiday, and they preferred to stay at home. In addition, due to road congestion at the Pasir Pekan roundabout, it was observed that there were a few traffic police to help control the traffic during the morning session from day 1 to day 5 (weekdays). The presence of traffic police was also observed at the Wakaf Bharu roundabout, since it was a public holiday, but they were not observed every day.

It can be concluded that the number of vehicles was influenced by various factors: weekdays/weekend, a public holiday and an event occurred during this time. During weekdays, there were many vehicles on the road, as people went to schools/workplaces, but the number lessened during the weekend.

The public holiday also played an important role, as the number of vehicles usually increased at the start and end of the public holiday due to people returning to their hometown at the start of the public holiday and returning to their respective homes at the end of the public holiday. In addition, the number of vehicles rose during the event as well but only during that time.

$PM_{2.5}$ and PM_{10} Concentrations at Each Roundabout

Graphs of $PM_{2.5}$ and PM_{10} concentrations at the three roundabouts are presented in Figures 5 to 7. For the roundabout at Pasir Pekan (Figure 2), the highest readings recorded for both PM concentrations were 336 ± 69.66 μg/m^3 for $PM_{2.5}$ and 352 ± 73.41 μg/m^3 for PM_{10}. For the roundabout at Wakaf Bharu (Figure 4), the highest readings recorded for both PM concentrations were 123 ± 34.51 μg/m^3 for $PM_{2.5}$ and 127 ± 35.72 μg/m^3 for PM_{10}. For the Tuan Padang roundabout at Kota Bharu (Figure 7), the highest readings recorded for both PM concentrations were 107 ± 27.48 μg/m^3 for $PM_{2.5}$ and 114 ± 29.22 μg/m^3 for PM_{10}. All of them occurred during the morning session.

Graphs of the mean concentrations of $PM_{2.5}$ and PM_{10} for all roundabouts are presented in Figure 8. All three roundabouts recorded the highest readings for both the mean $PM_{2.5}$ and PM_{10} concentrations during the morning session. For the roundabout at Pasir Pekan, the highest readings for both mean $PM_{2.5}$ and PM_{10} concentrations were 113 μg/m^3 and 126.86 μg/m^3, respectively. Similarly, both the Wakaf Bharu and Kota Bharu roundabouts recorded the highest readings, but only for mean $PM_{2.5}$ concentrations that exceeded the Malaysian Ambient Air Quality Standard (MAAQS): 90.86 μg/m^3 for the roundabout at Wakaf Bharu and 68.57 μg/m^3 for the Tuan Padang roundabout at Kota Bharu. The mean $PM_{2.5}$ and PM_{10} concentrations for both the afternoon and evening sessions did not exceed the limit.

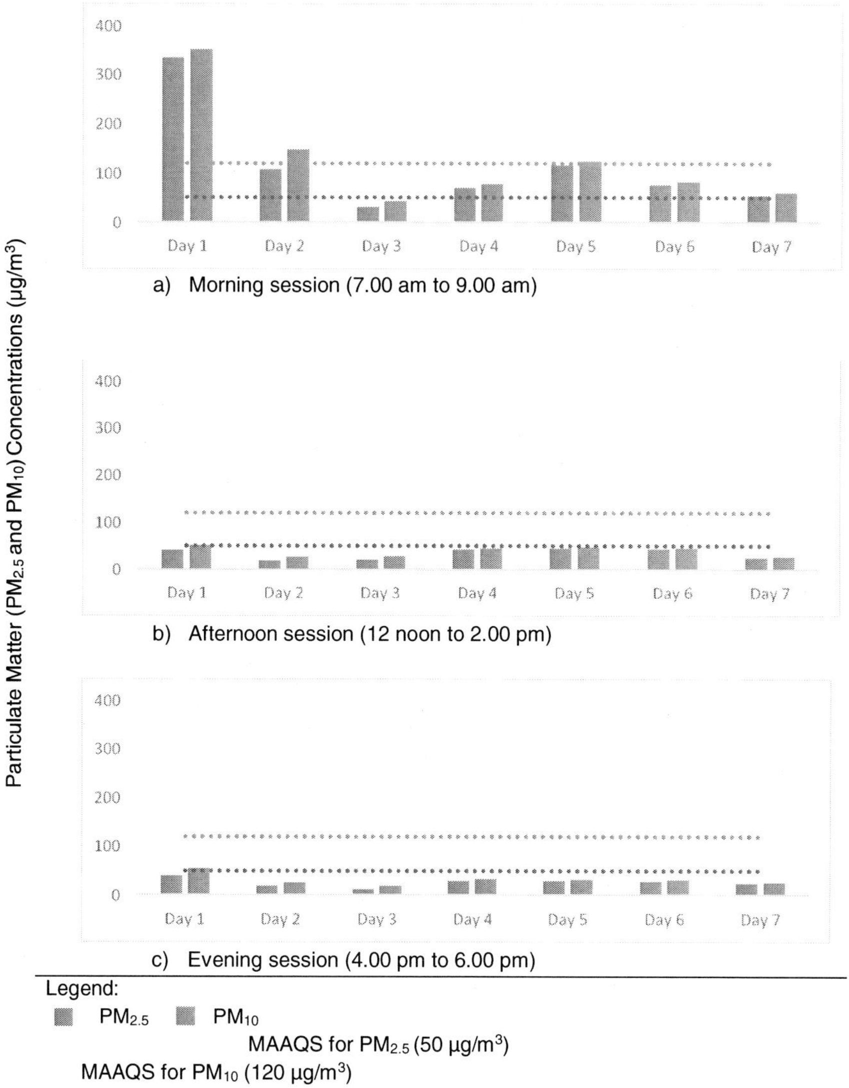

400
300
200
100
0
Day 1 Day 2 Day 3 Day 4 Day 5 Day 6 Day 7

a) Morning session (7.00 am to 9.00 am)

400
300
200
100
0
Day 1 Day 2 Day 3 Day 4 Day 5 Day 6 Day 7

b) Afternoon session (12 noon to 2.00 pm)

400
300
200
100
0
Day 1 Day 2 Day 3 Day 4 Day 5 Day 6 Day 7

c) Evening session (4.00 pm to 6.00 pm)

Legend:
$PM_{2.5}$ PM_{10}
MAAQS for $PM_{2.5}$ (50 $\mu g/m^3$)
MAAQS for PM_{10} (120 $\mu g/m^3$)

Figure 5. $PM_{2.5}$ and PM_{10} concentrations at the Pasir Pekan roundabout (R1).

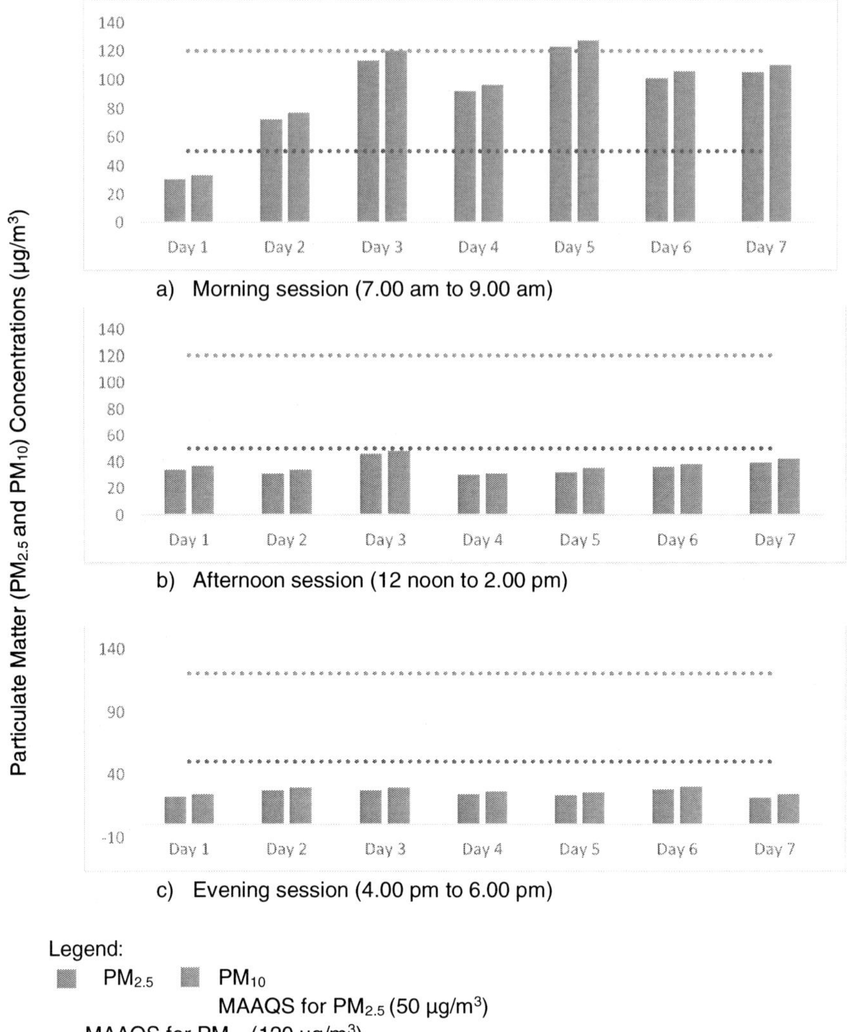

Figure 6. $PM_{2.5}$ and PM_{10} concentrations at the Wakaf Bharu roundabout (R2).

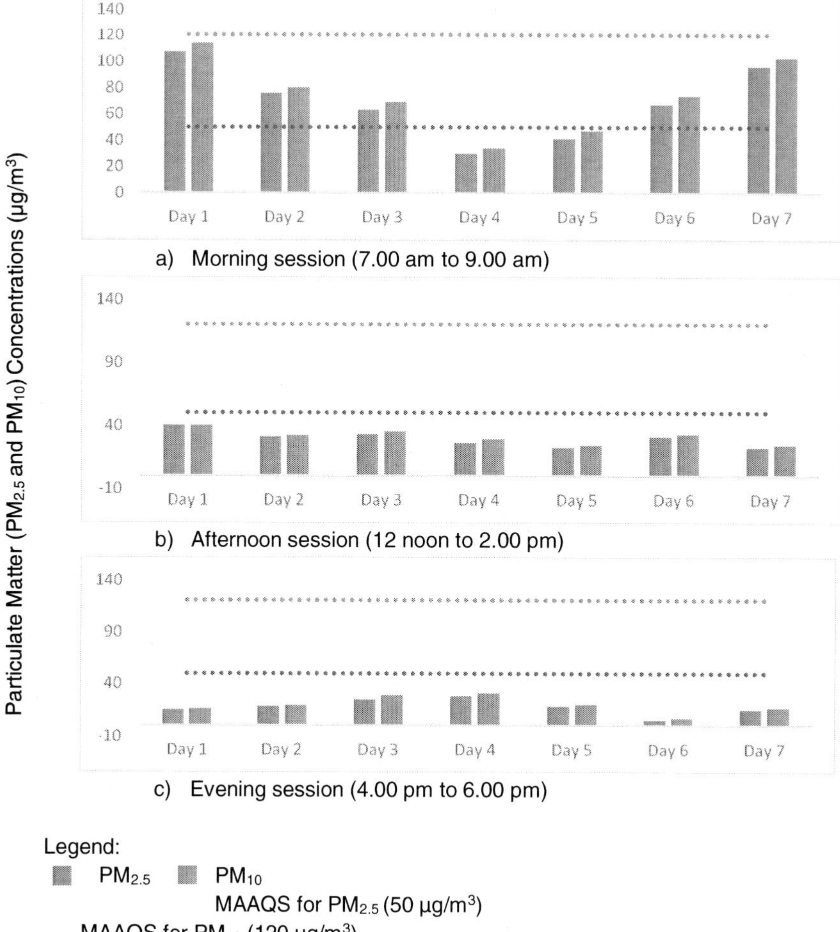

Figure 7. $PM_{2.5}$ and PM_{10} concentrations at the Tuan Padang roundabout, Kota Bharu (R3).

The mean particulate matter concentrations were the highest during the morning session (7.00 am-9.00 am), especially for $PM_{2.5}$. Since $PM_{2.5}$ has a diameter less than 2.5 µm, the low density makes it easier to disperse into the air than PM_{10}. It was believed that the highest mean PM concentration was due to people rushing to workplaces or schools in the morning. Furthermore, since the roundabouts were also connected to tourist attractions, such as Pasar Siti Khadijah, it was common for people to go to market in the morning, thus causing high readings. Furthermore, each roundabout also acts as the main road/connecting route to other places. Many vehicles use these roundabouts, thus explaining the high readings.

The results obtained were similar to those of San Martini et al. (2015), where $PM_{2.5}$ concentrations from selected provinces in China, namely, Beijing, Chengdu and Guangzhou, peaked between 7.00 a.m. and 8.00 a.m. and between 7.00 p.m. and 11.00 p.m. This implied that the morning session had the highest concentration, probably due to rush hour traffic at this time (San Martini et al. 2015). Nevertheless, the results obtained contradicted those of Wilkinson et al. (2013). They found that the ambient particulate matter was low in the morning (10.00 am), but the amount increased during the afternoon (12.00 pm), most certainly due to vehicle traffic intensifying during the lunch hours (Wilkinson et al. 2013).

Furthermore, there were only small and medium industries located within a 5 km radius from all roundabouts. These industries could not have had much effect on the measurements of the PM concentrations. Most major industries were located in the Pengkalan Chepa industrial area, which is more than 5 km from all roundabouts. Therefore, the $PM_{2.5}$ and PM_{10} readings were mostly measuring PM from vehicles but not from the industrial areas.

Inferential Statistics

Comparison of Particulate Matter ($PM_{2.5}$ and PM_{10}) Concentrations among Roundabouts

Table 3 shows the median $PM_{2.5}$ and PM_{10} concentrations for each roundabout. From the Kruskal-Wallis test, the p values for both $PM_{2.5}$ and PM_{10} were 0.345 and 0.238, respectively. The null hypothesis was not rejected, and there was no significant difference in $PM_{2.5}$ and PM_{10} concentrations among these three roundabouts. It can be stated that the amount of ambient PM during the monitoring process was similar; thus, there was no significant difference (refer to Table 3) due to the close proximity of each roundabout to each other.

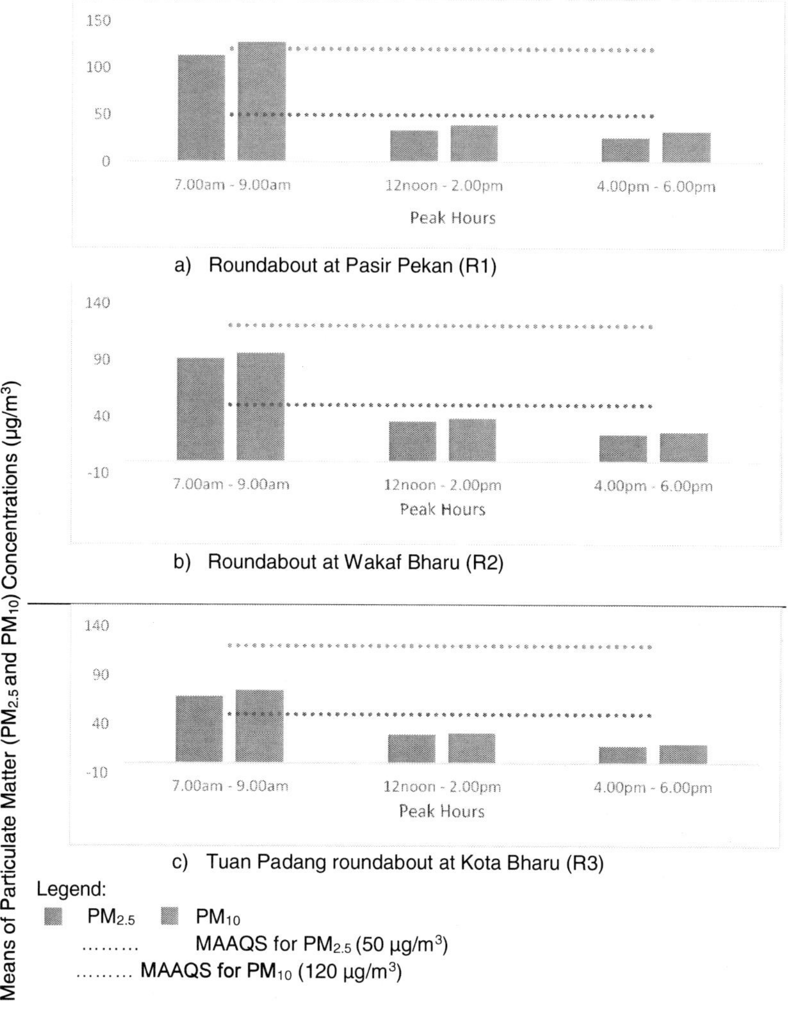

Figure 8. Mean $PM_{2.5}$ and PM_{10} concentrations at the Pasir Pekan roundabout, Wakaf Bharu roundabout and Tuan Padang roundabout at Kota Bharu.

Table 3. Median $PM_{2.5}$ and PM_{10} concentrations from the roundabouts

PM Concentrations	Roundabout	Median (IQR)	Z statistics	p value*
$PM_{2.5}$	R1	39.00 (38)	2.130 (2)	0.345
	R2	32.00 (55)		
	R3	30.00 (32)		
PM_{10}	R1	45.00 (42)	2.868 (2)	0.238
	R2	35.00 (58)		
	R3	32.00 (36)		

* Non-significant difference at $p > 0.05$, Statistical test - z statistics between three groups of roundabouts, Kruskal-Wallis test between three groups of roundabouts.

Number of Vehicles and Particulate Matter Concentrations

To determine the relationship between the number of vehicles and particulate matter concentration, it was important to identify whether the number of vehicles recorded had a normal distribution. Since it was not normal, the median was used instead. Then, Spearman correlation was used to determine the two variables mentioned above.

a) Difference in median of each type of vehicle. The Kruskal-Wallis test was used to find the difference in the median of each type of vehicle. The p value was less than 0.05, and there was a significant difference in the median number of vehicles (Table 4). In addition, *post hoc* analysis was performed by using the Mann-Whitney test to determine the difference for each pair. From the results, it could be said that all pairs were significantly different, except for the pair of lorries (144, IQR 160) and vans (140, IQR 86), with a p value of 4.014.
b) Correlation between the number of vehicles and particulate matter concentrations. The Spearman correlation test was used to determine the correlation between the number of vehicles and particulate matter ($PM_{2.5}$ and PM_{10}) concentrations (Table 5). In this analysis, the p value was less than 0.05. Thus, the null hypothesis was rejected. There was a weak and negative correlation between the number of vehicles and particulate matter

concentrations, with $r = -0.369$ and $p = 0.003$ for $PM_{2.5}$ and $r = -0.358$ and $p = 0.004$ for PM_{10}, with a corrected alpha of 0.01.

This correlation was believed to be influenced by rainfall during monitoring days. A study from Kwak et al. (2017) found that rainfall provided conflicting effects on the PM concentration, namely, direct and indirect effects. Rainfall has a direct washing effect that can remove pollutants to some extent. However, for indirect effects, rainfall is known to worsen air pollution by increasing traffic congestion and thus reducing vehicle speed. They concluded that rainfall significantly reduced the PM_{10} concentration by direct washing effects but also significantly increased the PM_{10} concentration by indirect pollution effects (Kwak et al. 2017).

Table 4. Median of vehicles according to type

Type of Vehicle	Median (IQR)	z statistics	*p* value*
Cars	7646 (2541)	338.274 (5)	<0.01
Motorcycles	3186 (1709)		
Buses	27 (54)		
Lorries	144 (160)		
Vans	140 (86)		
Others	10 (10)		

* Significant difference at $p < 0.05$, Statistical test - z statistics between six groups of vehicles, Kruskal-Wallis test between six groups of vehicles.

Table 5. Correlation between the number of vehicles and particulate matter ($PM_{2.5}$ and PM_{10}) concentrations

PM Concentrations	Number of Vehicle	
	r value	*p* value*
$PM_{2.5}$	-0.369	0.003
PM_{10}	-0.358	0.004

* Significant correlation at $p < 0.05$, statistical test - Spearman correlation test.

However, another study obtained different results from those in this study. They stated that the highest PM concentration occurred during morning peak traffic flow, between 8.00 am and 11.00 am, and another peak during evening peak traffic flow, between 5.00 pm and 9.00 pm. The PM levels were considerably lower between 12.00 noon and 4.00 pm. They concluded that the significant correlation between traffic volume count and $PM_{2.5}$ and PM_{10} concentrations indicated that traffic-related emissions were the main sources of emissions at the study site (Srimuruganandam and Shiva Nagendra 2010). A previous study stated that ambient particulate matter comes not only from exhaust emissions but also from non-exhaust emissions. Non-exhaust emissions typically come from abrasive sources, which include brake wear, tire wear and abrasion of the road surface (e.g., road wear and road tire) (Thorpe and Harrison 2008). In addition, the $PM_{2.5}$ and PM_{10} concentrations were similar between each roundabout, even though the mean number of vehicles was significantly lower at the Wakaf Bharu roundabout than at the other two roundabouts. This may be because people were stuck in road congestion; hence, with longer times on the road and thus leading to higher $PM_{2.5}$ and PM_{10} concentrations. It can be concluded that the relationship between the number of vehicles and particulate matter concentration depended on both exhaust and non-exhaust emissions.

CONCLUSION

In conclusion, the counting of vehicles in three selected roundabouts was divided into six types of vehicles. As expected, cars had the highest mean compared with the rest. The number of vehicles was dependent on weekdays/weekend, and a public holiday and an event occurred during the study period. The number of vehicles was higher on weekdays than on weekends. In addition, the Wakaf Bharu roundabout received the lowest mean number of vehicles (except for lorries) due to a public holiday. The distribution trend showed that $PM_{2.5}$ and PM_{10} concentrations were the highest in the morning (7.00 am-9.00 am) compared with the afternoon and evening. Both $PM_{2.5}$ and PM_{10} concentrations exceeded the MAAQS at the Pasir Pekan roundabout, whereas only $PM_{2.5}$ exceeded the MAAQS at the other two roundabouts. However, there was no significant difference in the

$PM_{2.5}$ and PM_{10} concentrations measured between the three roundabouts. It was concluded that all roundabouts had similar $PM_{2.5}$ and PM_{10} concentrations. High PM concentrations at the roundabouts were due to strategic location. In addition, PM readings were also influenced by road traffic emissions and geographical and meteorological parameters. For road traffic emissions, this study showed that the number of vehicles was inversely proportional to $PM_{2.5}$ and PM_{10} concentrations. This indicates that the $PM_{2.5}$ and PM_{10} concentrations increased with a decreased number of vehicles. Hence, the roundabout at Pasir Pekan is considered the most polluted, as it had the highest mean and median $PM_{2.5}$ and PM_{10} concentrations compared with the other two roundabouts, with a mean of 57.43 $\mu g/m^3$ and a median of 39.00 $\mu g/m^3$ for $PM_{2.5}$ and a mean of 65.86 $\mu g/m^3$ and a median of 45.00 $\mu g/m^3$ for PM_{10}.

ACKNOWLEDGMENTS

The authors would like to acknowledge the Universiti Sains Malaysia for supporting this study with Special Bridging Grant 2017, 604/PPSK/6316252. We would also like to thank all lecturers in the Environmental and Occupational Health programme, PPSK lecturers, all USM staff and friends who contributed directly or indirectly to this study.

REFERENCES

Badami, M. G. (2005). Transport and urban air pollution in India. *Environmental Management*, 36(2): 195-204.

Dominick, D., Juahir, H., Latif, M. T., Zain, S. M. & Aris, A. Z. (2012). Spatial assessment of air quality patterns in Malaysia using multivariate analysis. *Atmospheric Environment*, 60: 172-181.

European Environment Agency (EEA). (2010). *Emissions of primary $PM_{2.5}$ and PM_{10} particulate matter.* Retrieved on September 29,

2018, from https://www.eea.europa.eu/data-and-maps/indicators /emissions-of-primary-particles-and-5.

Gastaldi, M., Meneguzzer, C., Lucia, L. D. & Gecchele, G. (2014). Evaluation of air pollution impacts of a signal control to roundabout conversion using microsimulation. *Transportation Research Procedia*, 3: 1031-1040.

Google Maps. (2018). *Location of roundabout at Kota Bharu, Pasir Pekan and Wakaf Bharu*. Retrieved on December 20, 2018, from https://www.google.com/maps.

Gulia, S., Shiva Nagendra, S. M., Khare, M. & Khanna, I. (2015). Urban air quality management - A review. *Atmospheric Pollution Research*, 6(2): 286-304.

Kanabkaew, T., Nookongbut, P. & Soodjai, P. (2013). Preliminary assessment of particulate matter air quality associated with traffic emissions in Nakhon Si Thammarat, Thailand. *Procedia Engineering*, 53: 179-184.

Kwak, H. Y., Ko, J., Lee, S. & Joh, C. H. (2017). Identifying the correlation between rainfall, traffic flow performance and air pollution concentration in Seoul using a path analysis. *Transportation Research Procedia*, 25:3552-3563.

Mandavilli, S., Rys, M. J. & Russell, E. R. (2008). Environmental impact of modern roundabouts. *International Journal of Industrial Ergonomics,* 38(2): 135-142.

San Martini, F. M., Hasenkopf, C. A. & Roberts, D. C. (2015). Statistical analysis of $PM_{2.5}$ observations from diplomatic facilities in China. *Atmospheric Environment,* 110:174-185.

Srimuruganandam, B. & Shiva Nagendra, S. M. (2010). Analysis and interpretation of particulate matter - PM_{10}, $PM_{2.5}$ and PM_1 emissions from the heterogeneous traffic near an urban roadway. *Atmospheric Pollution Research*, 1(3): 184-194.

The Population and Housing Census of Malaysia. (2010). T*otal population by ethnic group, district/mukim and state in Malaysia in 2010.* Retrieved on March 2, 2019, from www.mdtumpat.gov. my/sites/default/files/mukim_kelantan.pdf.

Thorpe, A., & Harrison, R. M. (2008). Sources and properties of non-exhaust particulate matter from road traffic: A review. *Science of The Total Environment*, 400(1-3): 270-282.

United States Environmental Protection Agency (USEPA). (2018). *Particulate matter (PM) basics.* Retrieved on September 23, 2018, from https://www.epa.gov/pm-pollution/particulate-matter-pm-basics#PM.

Wang, H., Fu, L., Zhou, Y., Du, X. & Ge, W. (2010). Trends in vehicular emissions in China's mega cities from 1995 to 2005. *Environmental Pollution*, 158(2): 394-400.

Wilkinson, K. E., Lundkvist, J., Netrval, J., Eriksson, M., Seisenbaeva, G. A. & Kessler, V. G. (2013). Space and time resolved monitoring of airborne particulate matter in proximity of a traffic roundabout in Sweden. *Environmental Pollution*, 182: 364-370.

World Health Organization (WHO). (2018a). *Exposure to ambient air pollution from particulate matter for 2016*. Retrieved on March 13, 2019, from https://www.who.int/airpollution/data/AAP_ exposure_ Apr2018_final.pdf?ua=1.

World Health Organization (WHO). (2018b). *What is air pollution?* Retrieved on April 4, 2019, from http://www.searo.who.int/topics/ air_pollution/what-is-air-pollution.pdf?ua=1.

World Health Organization (WHO). (2018c). *Ambient (outdoor) air quality and health.* Retrieved on September 23, 2018, from http://www.who.int/news-room/factsheets/detail/ambient-(outdoor)-air-quality-and-health.

Wu, J. Z., Ge, D. D., Zhou, L. F., Hou, L. Y., Zhou, Y. & Li, Q. Y. (2018). Effects of particulate matter on allergic respiratory diseases. *Chronic Diseases and Translational Medicine*, 4(2): 95-102.

In: The Influence of Ecosystem Services ... ISBN: 978-1-53619-977-2
Editors: Hasmah Abdullah et al.

Chapter 5

DETECTION OF HEAVY METALS IN WASTEWATER DISCHARGE FROM INDUSTRIAL, AGRICULTURAL AND DOMESTIC HOUSEHOLD AREAS AROUND KOTA BHARU, KELANTAN

Nurasmat M. S.[1,*], Nur Fatien M. S.[1], Nur Izzah Azira A. R.[1], Norshahidatul Akmar M. S.[2], Wan Nazwanie W. A.[3], Nor Hakimin A.[4] and Zamani A. H.[5]

[1]School of Health Sciences, Health Campus, Universiti Sains Malaysia, Kubang Kerian, Kelantan, Malaysia
[2]Faculty of Applied Sciences, Universiti Teknologi MARA Pahang, Jengka, Pahang, Malaysia
[3]School of Chemical Sciences, Universiti Sains Malaysia, Pulau Pinang, Malaysia

[*] Corresponding Author's E-mail: nurasmatms@usm.my.

[4] Advanced Materials Research Cluster (AMRC), Faculty of Bioengineering and Technology, Universiti Malaysia Kelantan (UMK), Jeli, Kelantan, Malaysia
[5] Faculty of Industrial Sciences and Technology, College of Computing and Applied Science, Universiti Malaysia Pahang, Gambang, Kuantan Pahang, Malaysia

ABSTRACT

Heavy metal pollution of wastewater is a serious issue for both humans and the environment. The concentration of heavy metals such as lead (Pb), cadmium (Cd), chromium (Cr), zinc (Zc), and copper (Cu) in wastewater discharged from industrial, agriculture, and domestic household areas were investigated in this study. The elements were analysed using atomic absorption spectroscopy (AAS) after the wastewater sample was digested using the wet acid digestion method. The highest concentration of Cd and Cu were found in the industrial area, while the highest concentration of Pb, Zn, and Cr were detected in the agricultural farm area. Cd effluent in wastewater of industrial sites, agriculture farms and domestic households found to be exceeded the permissible level. These results revealed that proper wastewater treatment and monitoring are vitally crucial for the three areas.

Keywords: heavy metals, wastewater, atomic absorption spectroscopy, acid digestion method

INTRODUCTION

Currently, many land areas have been developed for housing, agricultural farms and industries. The increase in urbanization and industrialization has caused serious pollution to the environment, particularly to water, due to the increase in wastewater discharge (Yuce et al., 2006; Suratman et al., 2009). Wastewater may contain heavy metals that potentially pollute the water surface due to their nonbiodegradable properties (Parvin et al., 2019). Consequently, the accumulation of heavy metals in water bodies may have carcinogenic effects on humans and deleterious effects on the environment (He et al., 2013; Parvin et al., 2019).

Generally, the heavy metal content within wastewater may originate from the discharge of industrial effluent, runoff of pesticides or fertilizers from agricultural farms, and domestic sewage discharge (Bhuiyan et al., 2015; Shazili et al., 2006). Due to the high awareness of the public regarding the detrimental effects of heavy metals on health and the environment, several countries have implemented regulations to control and monitor the permissible levels of heavy metals discharged in effluent. Based on the United States Environmental Protection Agency (EPA), the permissible cadmium (Cd), lead (Pb), zinc (Zn) and chromium (Cr) contents within wastewater were 0.01 mg/L, 0.05 mg/L, 5 mg/L and 0.05 mg/L, respectively (Yayintas et al., 2007). According to the Malaysia Standard of Effluence Discharge, the permissible concentrations of Cd, Pb, Cu, Zn and Cr in effluent were 0.02 mg/L, 0.5 mg/L, 1.0 mg/L, 1.0 mg/L and 1.0 mg/L, respectively (Malaysia Enviromental Quality 2000).

Elemental information of water samples from rivers has been explored by several authors using various instrumental techniques, including inductively coupled plasma-optical emission spectroscopy (Poh et al., 2008; Suratman et al., 2009), inductively coupled plasma-mass spectrometry (Khalik et al., 2013) and graphite furnace atomic absorption spectroscopy (Poh et al., 2008). These studies have undoubtedly provided significant contributions in terms of heavy metal analysis of water; however, they have mainly focused on rivers. Nonetheless, limited information can be found on the concentrations of heavy metals in wastewater discharged into drains. Therefore, a proper study based on the heavy metals in wastewater needs to be conducted. In this study, the chemical properties of heavy metals in wastewater will be analysed using atomic absorption spectroscopy (AAS).

METHODOLOGY

Sampling Procedure

Wastewater sampling was conducted at three different sampling sites, i.e., an industrial site, agricultural farm and domestic household located in the area of Pengkalan Chepa, Kelantan, Malaysia. Each sampling site was located at four sampling points, with a distance of approximately 500 m between each point. Table 1 lists the wastewater samples based on their sampling sites. Wastewater samples were collected in sterile plastic containers. Upon collection, wastewater samples were preserved by adding 1.5 mL of concentrated nitric acid (HNO_3).

Table 1. Sampling sites of wastewater samples

Wastewater Samples	Sampling Sites
A1	Industrial Site, Ain Medicare Factory, Pengkalan Chepa 2, Kelantan
A2	
A3	
A4	
B1	Agricultural Farm, Fruit Farm Pengkalan Chepa, Kelantan
B2	
B3	
B4	
C1	Domestic Household, Kompleks Perumahan PDRM, Pengkalan Chepa, Kelantan
C2	
C3	
C4	

Chemicals and Reagents

Trace metal grade 65% nitric acid (HNO_3) was obtained from HMbG Chemicals. The element stock solutions, cadmium (Cd), lead (Pb), copper

(Cu), zinc (Zn), and chromium (Cr), were obtained from Merck, with each concentration being 1000 mg/L.

Preparation of Calibration Standards

Each of the element standard solution (Cd, Pb, Zn, Cr, and Cu) was prepared with five different concentrations by serial dilution from 1000 mg/L of a standard stock solution. Cr and Cu were diluted to obtain a calibration range from 0.3125 mg/L to 5.0 mg/L. Moreover, Cd, Pb and Zn were diluted from a 1000 mg/L stock solution into ranges from 0.125 mg/L to 2.0 mg/L, 1.25 mg/L to 20.0 mg/L and 0.0625 mg/L to 1.0 mg/L, respectively.

Sample Preparation

Sample preparation was conducted using wet acid digestion. Fifty millilitres of the wastewater sample were measured and transferred into a cleaned 100 mL beaker and subsequently added to 5 mL of concentrated HNO_3. The beaker was closed with watch glass and heated on a hot plate to evaporate the solvent in the beaker until only 10 mL to 20 mL remained. After the first stage of the digestion process, another 2.5 mL of HNO_3 was added to the beaker, and the beaker was covered by watch glass before being subjected to hot plate heating to completely digest the sample. The digested sample was filtered using Whatman filter paper and diluted with deionized water to 100 mL. A reagent blank solution was also used as a negative control.

Instrumental Analysis

A Perkin Elmer AAnalyst800 Atomic Absorption Spectroscopy (AAS) available in the Analytical Laboratory, Universiti Sains Malaysia, was utilized in this study. Prior to sample analysis, the instrument was calibrated with all target elements at five calibration points. Upon generation of the calibration curve, the sample was aspirated into the AAS system. For quantitative analysis, the correlation coefficients (r) of the calibration curves for Cu, Cd, Zn, Cr and Pb were 0.9995, 0.9782, 0.9924, 0.9998, and 1.0000, respectively.

Data Analysis

Data interpretation was performed using Microsoft Excel. The concentrations of heavy metals (Cu, Cd, Pb, Cr, and Zn) among different wastewater samples obtained from different sampling points within the same sampling site were compared and reported in this study. Moreover, the concentrations of heavy metals in wastewater samples from different sampling sites were compared with the permissible level set by Malaysia Environmental Quality (Sewage and Industrial Effluents) Regulations.

Results and Discussion

Heavy Metal Analysis for Different Sampling Points

The heavy metal (Cd, Pb, Cu, Cr, and Zn) contents of the wastewater at different sampling points were determined, and the results are given in Table 2.

For the industrial site, Cr had the highest concentration (0.019 mg/L) at point A1. The highest concentrations of Cd and Cu were located at point A3, with concentrations of 1.025 and 0.481 mg/L, respectively, while the highest concentrations of Pb and Zn were found at point A4, with concentrations of 0.058 and 0.549 mg/L, respectively. For the agricultural site, Cd, Cr, and Zn had the highest concentrations, 1.226 mg/L, 1.787 mg/L, and 0.589 mg/L, at points B4, B1, and B3, respectively. Nevertheless, Pb and Cu exhibited the highest concentrations at point B2, with concentrations of 0.12 and 0.034 mg/L, respectively. For the household site, the highest concentrations of Cd and Cu were shown at C2 and C1, with concentrations of 0.998 mg/L and 0.141 mg/L, respectively, while the highest concentrations of Pb and Zn were found at point C4, with concentrations of 0.118 and 0.598 mg/L, respectively. The various concentrations of wastewater heavy metals at different sampling points may have been due to the historical heavy metal residues that had accumulated in drainage sediment at each sampling point (Poh et al., 2008).

Table 2. Concentration of heavy metals at different sampling points

Sampling Points	**Sampling Site**	**Heavy Metal Concentration (mg/L)**				
		Cd	Pb	Cu	Cr	Zn
A1	Industrial	1.008	0.048	0.013	0.019	0.406
A2		1.009	ND	0.013	0.011	0.361
A3		1.025	0.014	0.481	ND	0.506
A4		1.012	0.058	0.234	ND	0.549
B1	Agriculture	1.053	0.094	0.009	1.787	0.54
B2		0.721	0.12	0.034	1.702	0.512
B3		1.018	0.102	0.015	ND	0.589
B4		1.226	0.11	0.005	1.668	0.572
C1	Household	0.988	0.038	0.141	ND	0.54
C2		0.998	0.038	0.067	ND	0.063
C3		0.794	0.026	0.008	ND	0.569
C4		0.903	0.118	0.007	ND	0.598

*ND = not detected.

Heavy Metal Analysis for Different Sampling Sites

The concentrations of each heavy metal in the wastewater samples at three sampling sites (industrial, agricultural and household) are given in Table 3. The results are reported as the mean for each element.

Table 3. Concentration of heavy metal sampling in three different areas

Sampling Sites	Concentration of Heavy Metals (mg/L)				
	Cd	Pb	Cu	Zn	Cr
Industrial	1.014	0.030	0.185	0.456	0.008
Agricultural Farm	1.005	0.107	0.016	0.553	1.290
Domestic Household	0.921	0.055	0.056	0.443	ND

Based on Table 3, the highest concentrations of Cd and Cu were detected at industrial sites, with concentrations of 1.014 and 0.185 mg/L, respectively. The highest concentrations of Cd and Cu in the industrial area may have been due to industrial manufacturing activities, such as electroplating and preparation of dye and pigments, as well as the discharge of municipal wastewater (Shazili et al., 2006; Kaushik et al., 2009; Bhuiyan et al., 2015;). Pb, Zn, and Cr had the highest concentrations at the agricultural farm, with concentrations of 0.107 mg/L, 0.553 mg/L, and 1.290 mg/L, respectively. The highest concentrations of Pb, Zn, and Cr in the wastewater in the agricultural farm area were contributed by the runoff of fertilizer or pesticide from agricultural land into the wastewater (Shazili et al., 2006; Bhuiyan et al., 2015). Moreover, the presence of Pb and Zn contents in wastewater within the agricultural farm area may have potentially been contributed by petrol with lead and tire wear from high vehicle traffic (Shazili et al., 2006; Khalik et al., 2013).

Comparison of the Concentration of Heavy Metals in Wastewater with the Malaysia Standard Effluent Discharge

The accumulation of heavy metals in rivers is mainly due to wastewater discharge from anthropogenic activities. Due to the toxic effects of heavy metals on public health and the environment, some countries have set permissible levels for heavy metal contents in wastewater. Table 4 shows the Malaysia Standard Effluence Discharge in accordance with Malaysia Environmental Quality, 2000.

Table 4. Permissible level of wastewater heavy metals based on the Malaysia Standard Effluence Discharge

Heavy Metals	Permissible Level (mg/L)
Cadmium	0.02
Chromium	1.0
Copper	1.0
Zinc	1.0
Lead	0.5

According to the results shown in Table 4, the concentrations of Cd in the industrial area, domestic household and agricultural farm exceeded the permissible level set by the Malaysia Environmental Regulation. Cadmium is the most toxic heavy metal to the environment and can cause detrimental health effects to humans, such as renal dysfunction, osteoporosis, cardiovascular disease, and liver damage (Fan et al., 2018). Therefore, the mitigation and monitoring of the heavy metal content in wastewater within industrial, agricultural and domestic household areas are necessary to ensure that the concentration of heavy metal contents in wastewater is below the permissible level set by Malaysia Environmental Quality.

CONCLUSION

This study successfully determined the levels of cadmium, chromium, lead, copper and zinc in wastewater. Cd and Cu had the highest concentrations at the industrial site, with concentrations of 1.014 mg/L and 0.185 mg/L, respectively, while Pb, Zn and Cr had the highest concentrations at the agricultural farm, with concentrations of 0.107 mg/L, 0.553 mg/L, and 1.290 mg/L, respectively. Compared with the Malaysia Standard Effluence Discharge, the cadmium concentrations in wastewater from the industrial area, agricultural farm and domestic household exceeded the permissible level. Therefore, proper wastewater treatment and monitoring must be conducted within the Pengkalan Chepa area.

Acknowledgments

The authors greatly appreciate the instrumental and financial support received from the Universiti Sains Malaysia (USM).

References

Bhuiyan, M. A. H., Dampare, S.B., Islam, M. A. & Suzuki, S. (2015). Source apportionment and pollution evaluation of heavy metals in water and sediments of Buriganga River, Bangladesh, using multivariate analysis and pollution evaluation indices. *Environmental Monitoring and Assessment*, 187(1): 4075.

Fan, R., Hu, P. C., Wang, Y., Lin, H. Y., Su, K., Feng, X. S., Wei, L. & Yang, F. (2018). Betulinic acid protects mice from cadmium chloride-induced toxicity by inhibiting cadmium-induced apoptosis in kidney and liver. *Toxicology Letters*, 299: 56-66.

He, B., Yun, Z., Shi, J. & Jiang, G. (2013). Research progress of heavy metal pollution in China: sources, analytical methods, status, and toxicity. *Chinese Science Bulletin*, 58(2): 134-140.

Kaushik, A., Kansal, A., Kumari, S. & Kaushik, C.P. (2009). Heavy metal contamination of river Yamuna, Haryana, India: assessment by metal enrichment factor of the sediments. *Journal of Hazardous Materials*, 164(1): 265-270.

Khalik, W. M. A. W. M., Abdullah, M. P. & Sani, N. A. A. (2013). Preliminary studies on sediment characteristics and metals contaminants of Temenggor lake, Malaysia. *Journal of Sustainability Science and Management*, 8(1): 80-86.

Malaysia Environmental Quality. (2000). *Malaysia Standard of Effluence Discharge, 2000. Retrieved on February 10, 2020 from Malaysia Department of Environment* http://www.doe.gov.my/index.php?option=com_content&task=blogsection&id=8&Itemid=54&lang=en.

Parvin, F., Rikta, S. Y. & Tareq, S. M. (2019). Application of nanomaterials for the removal of heavy metal from wastewater. In *Nanotechnology in water and wastewater treatment* (pp. 137-157). Elsevier.

Poh, S. C., Suratman, S., Chew, C. K., Shazili, N. A. M. & Tahir, N. M. (2008). Metal geochemistry of Nerus River, Terengganu. *Malaysian Journal of Analytical Sciences,* 12(3): 593-599.

Shazili, N. A M., Yunus, K., Ahmad, A. S., Abdullah, N. & Rashid, M. K. A. (2006). Heavy metal pollution status in the Malaysian aquatic environment. *Aquatic Ecosystem Health & Management*, 9(2): 137-145.

Suratman, S., Hang, H. C., Shazili, N. A. M. & Tahir, N. M. (2009). A preliminary study of the distribution of selected trace metals in the Besut River basin, Terengganu, Malaysia. *Bulletin of Environmental Contamination and Toxicology,* 82(1): 16-19.

Yayintas, O.T., Yılmaz, S., Turkoglu, M., & Dilgin, Y. (2007). Determination of heavy metal pollution with environmental physicochemical parameters in wastewater of Kocabas Stream (Biga, Canakkale, Turkey) by ICP-AES. *Environmental Monitoring and Assessment*, 127(1-3): 389-397.

Yuce, G., Pinarbasi, A., Ozcelik, S. & Ugurluoglu, D. (2006). Soil and water pollution derived from anthropogenic activities in the Porsuk River Basin, Turkey. *Environmental Geology*, 49(3): 359-375.

In: The Influence of Ecosystem Services … ISBN: 978-1-53619-977-2
Editors: Hasmah Abdullah et al.

Chapter 6

EFFECT OF ABSORPTION OF TERRESTRIAL GAMMA RADIATION TO THE BODY ORGANS IN THE ENVIRONMENT OF MICHIKA AREA NORTH-EASTERN NIGERIA

Gabdo H. T.[1,*], Idi B. Y.[2] and Inuwa I. A.[1]
[1]Department of Physics, Federal College of Education Yola, Adamawa State, Nigeria
[2]Department of Physics, Federal University Dutse, Jigawa State, Nigeria

ABSTRACT

Measurements for Terrestrial Gamma Radiation (TGR) dose rate were made in Michika area of north-eastern Nigeria using aeromagnetic survey in 6983 locations with a range value from 15 nGy h-1 to 324 nGy h^{-1} and the mean value of 115 ± 0.5 nGy h^{-1}. The outdoor annual effective dose is found to be 0.141 mSv $y.^{-1}$. The computed lifetime effective dose, cancer risk and the lifetime cancer risk from outdoors exposure for each

[*] Corresponding Author's E-mail: tghamman2@gmail.com.

person living in the Michika area is found to be 7.614 mSv, 8.21×10^{-3} and 4.43×10^{-1} respectively. These values are two times higher than the world outdoor average values of 4.9 mSv, 4.07×10^{-3}, and 2.85×10^{-1} respectively. The effective doses are possibly due to inhalation of gamma radiation targeting the internal organs like the gonads (testes or ovaries), lung, liver, and the skin of the body with values of 0.023μSv, 0.014μSv, 0.006μSv and 0.001μSv respectively. An Isodose map of the study area was presented to depict areas of enhancing TGR.

Keywords: Terrestrial Gamma Radiation (TGR) dose rate, annual effective dose, Tissue weighing factor, cancer risk.

INTRODUCTION

Over the years, there have been various worldwide studies that have quantified the level of terrestrial gamma radiation (TGR) in the environment. Previous studies have shown different absorption rates of TGR in the body. However, this bodily effect is generally a result of the tissue- or organ-specific weighting factor (W_T), which accounts for the variations in the risk to individual organs or tissues that constitute the body. In general, measurements of the terrestrial gamma radiation (TGR) dose rate in the environment are important for determining the level of exposure to radiation by persons living within an area. According to the UNSCEAR 2000, exposure to radiation can damage living cells, causing death in some of them and modifying others. The organs and tissues of the body lose a certain number of cells after they are exposed to radiation that exceeds a threshold level.

Radiation exposure may also lead to certain diseases, such as leukaemia, or cancers of many organs, such as the lung, breast and thyroid gland. However, radiation exposure due to a low dose (which is approximately similar to the global average value) would not elicit significant chances of developing an attributable cancer. However, continuous exposure over time can overturn the level of the attributable cancer.

In this study, terrestrial gamma radiation (TGR) dose rate measurements were used to estimate the radiation hazard indices that are associated with exposure to TGR in the Michika local government area of Adamawa state, north-eastern Nigeria. This provides insights into the effect of this radiation on body organs due to inhalation.

STUDY AREA

The study area (the Michika Local Government) is one of the 21 Local Government Areas of Adamawa State, north eastern Nigeria (Figure 1). The local government consists of 8 districts and 16 wards (Figure 2) and has a population of 179,460 individuals (2011 projection) with an area of 967 km^2 and a density of 186/km^2. The area has longitude coordinates of 13°19'E to 13°25'E and latitude coordinates of 10°32'N to 10°41'N. The area is bordered to the east by the Republic of Cameroon, to the south by the Mubi local government area of Adamawa State, to the west by the Askira Uba local government area of Borno State and to the north by the Madagali local government area (Figure 2). The area is hilly in the eastern part and relatively flat in the west; additionally, despite the hilly nature of some parts of the area, there are good road networks, foot paths and tracks (Nur A and Ayuni 2011).

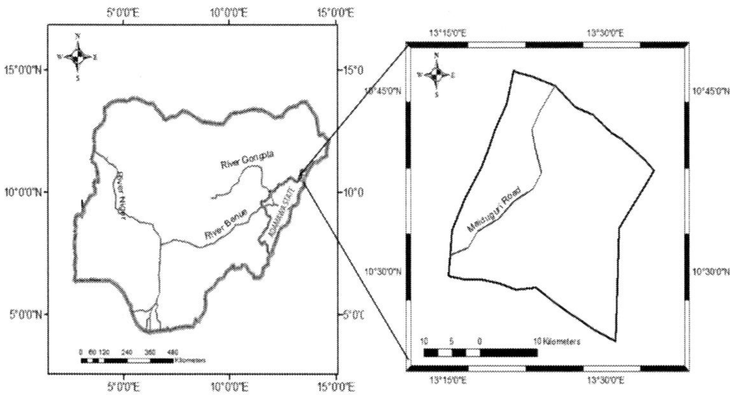

Figure 1. Map of Nigeria showing the location of the Michika local government.

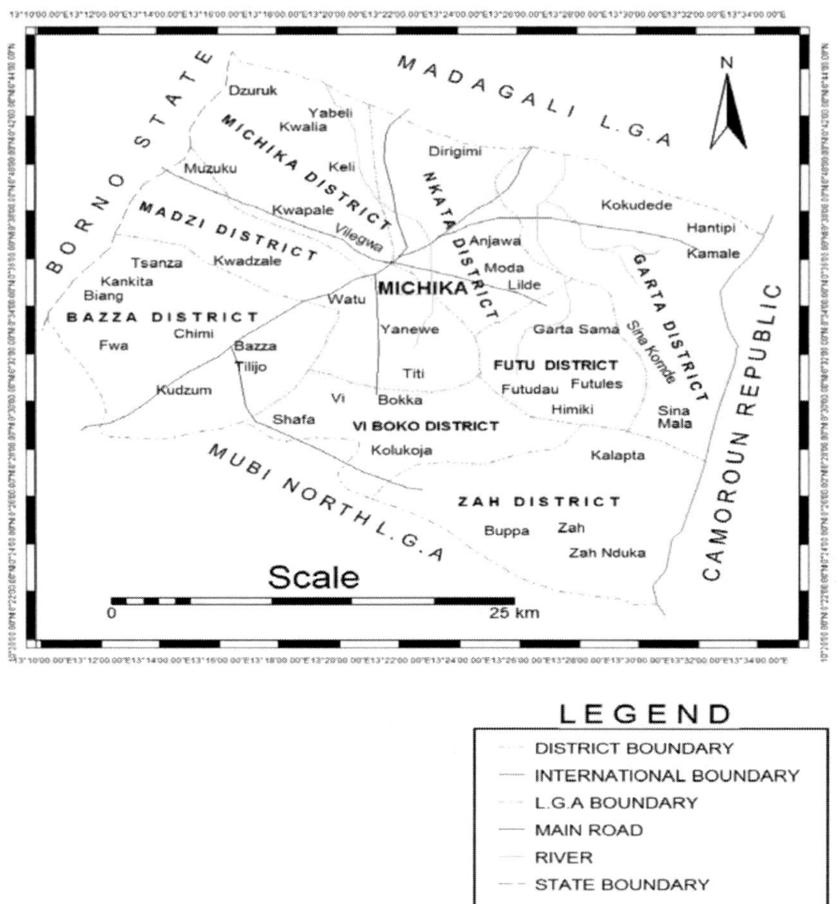

Figure 2. District map of the Michika local government area (James et al., 2015).

Materials and Methods

Airborne Survey

This study used airborne survey data obtained from the airborne survey of the Northeast project Block D1 (Figure 3). This high-resolution geophysical survey involving magnetic, radiometric and limited electromagnetic surveys was performed with the use of fugro airborne surveys between 2003 and 2010 on the behaviour of the geological survey agency of Nigeria. This agency survey was aimed at assisting and promoting mineral exploration in the country. The survey used radiometric data acquisition equipment GR-820-3 with radiometric crystals-GPX1024/256. Additionally, the flight path navigation equipment NOVATEL3151R/Omnister RTDGPS was used. The flight line spacing was 200-500 m, with a data recording interval of 1 sec.

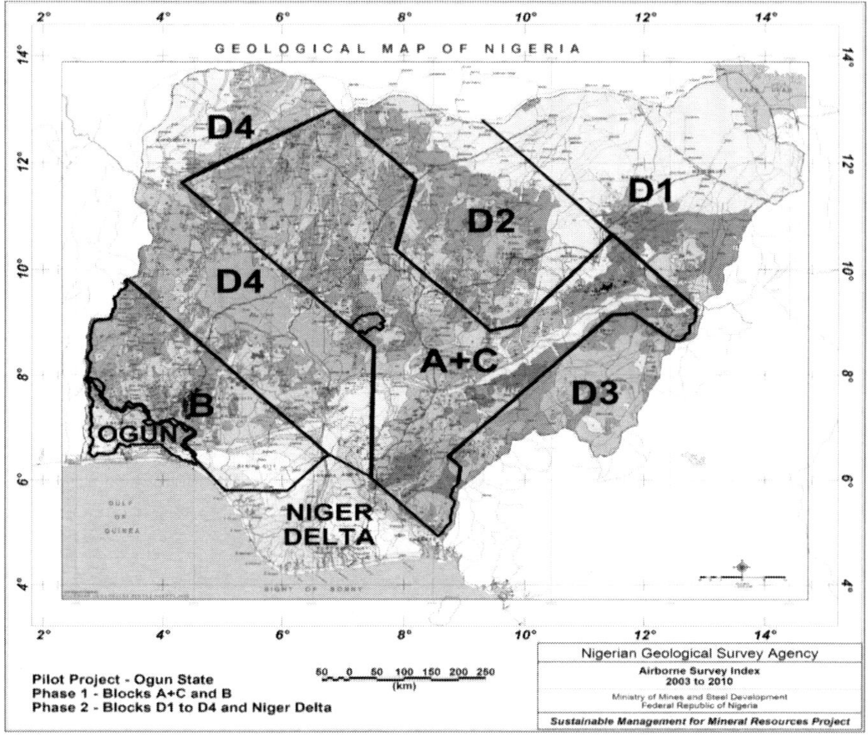

Figure 3. Index map between 2003 and 2010 (SEG Denver 2010).

The result of the survey in Block D1, which comprised the study area, shows a different range of TGR dose rates over the 6,983 locations that were measured in the area. These differences in range may be due to the presence of different geological formations (Figure 4) and soil types (Figure 5) in the area that influence the background radiation.

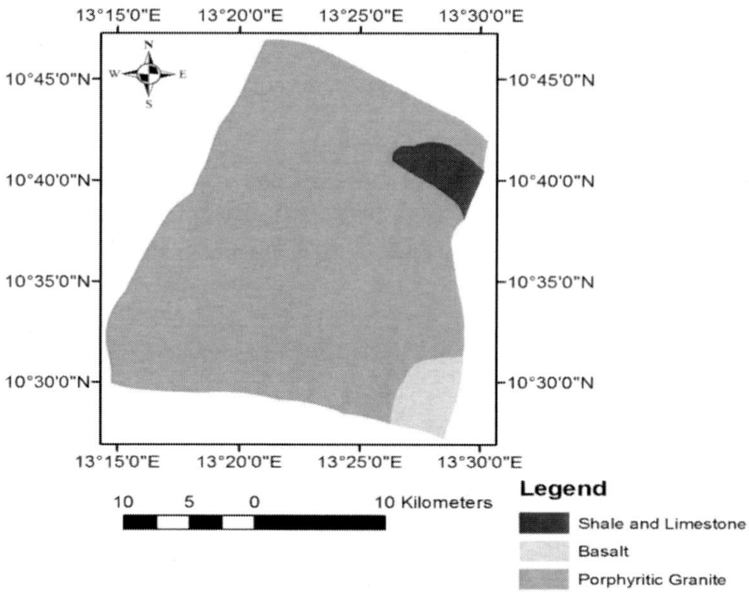

Figure 4. Geological map of the study area.

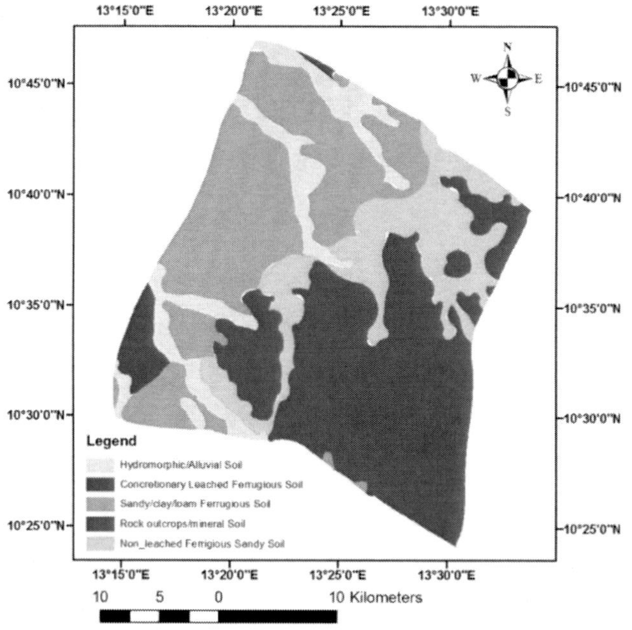

Figure 5. Soil map of the study area.

RESULTS AND DISCUSSION

Terrestrial radiation dose rates were measured from 6,983 outdoor locations (Figure 6), with values ranging from 15 nGy h^{-1} to 324 nGy h^{-1} and a mean value of 115 ± 0.5 nGy h^{-1} (Table 1). This value is two times the world average of 59 nGy h^{-1} (UNSCEAR 2000). Enhanced activities ranging between 200-300 nGy h^{--1} were observed in the Vboko, Futu, Madzi and Zah districts. The highest measurement of 324 nGy h^{-1} was observed in the Vboko district, with the lowest value of 15 nGy h^{-1} measured in the Zah district (Table 2).

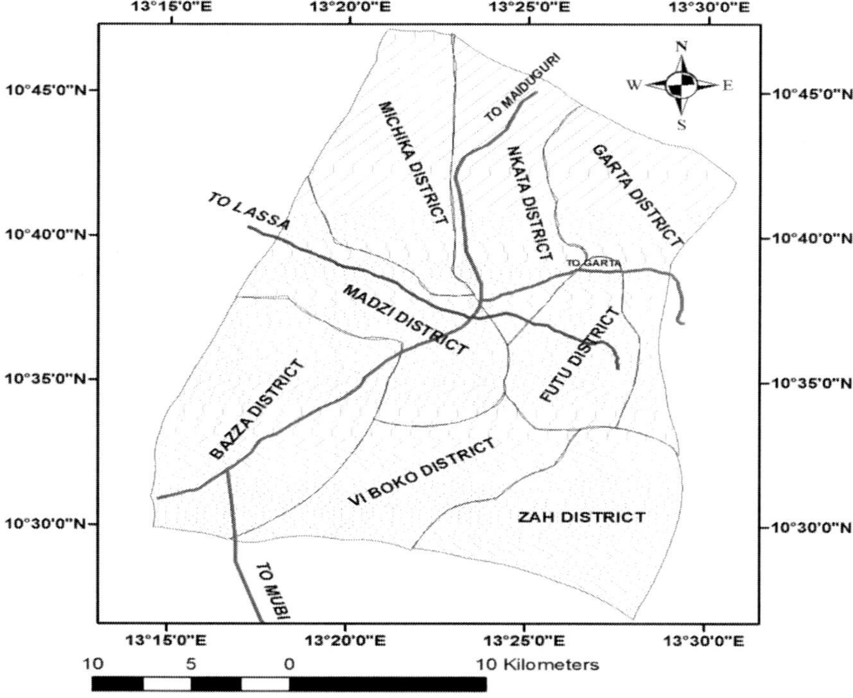

Figure 6. Dose rate locations along the flight path.

The descriptive statistics of the terrestrial gamma radiation dose rates for the study area are shown in Table 1. Table 2 shows the statistics for the eight (8) districts in the study area.

Table 1. Descriptive statistics of TGR measurements in the study area

			Statistic	Std. Error
DOSE_RATE	Mean		115.09	0.545
	95% Confidence Interval for Mean	Lower Bound	114.03	
		Upper Bound	116.16	
	5% Trimmed Mean		112.63	
	Median		111.84	
	Variance		2067.670	
	Std. Deviation		45.472	
	Minimum		15	
	Maximum		324	
	Range		310	
	Interquartile Range		52	
	Skewness		0.819	0.029
	Kurtosis		1.139	0.059

Table 2. Descriptive statistics of the TGR in the districts of the study area

	Minimum	**Maximum**	**Mean**		**Std. Deviation**
	Statistic	Statistic	Statistic	Std. Error	Statistic
VBOKO	26	324	127.30	2.650	72.473
BAZZA	42	183	100.80	0.764	27.058
FUTU	24	297	122.21	3.149	73.572
GARTA	46	210	114.64	0.901	25.928
NKATA	24	178	102.91	0.888	25.902
MADZI	20	273	97.72	1.450	42.628
MICHIKA	45	175	116.66	0.719	21.401
ZAH	15	295	148.66	1.435	43.015

Figure 7 shows the frequency distribution of the gamma dose rate measurements in the study area, and Figure 8 shows the P-P and Q-Q plots of the data. These figures show a near normal distribution of the data.

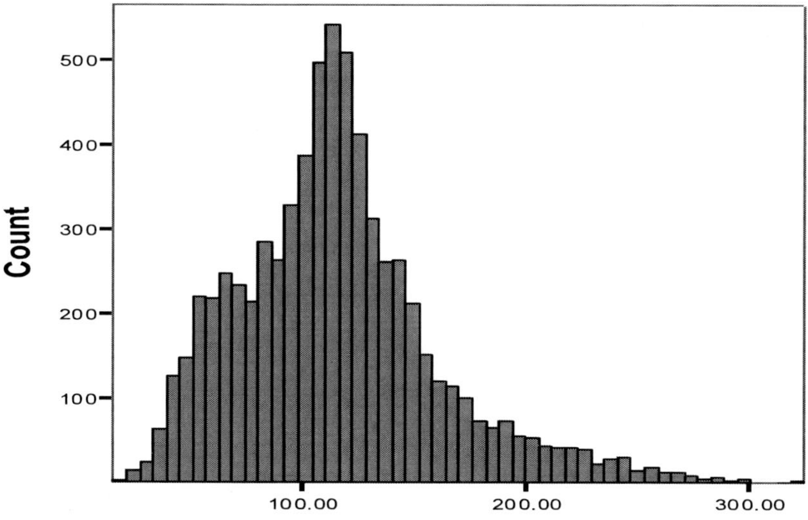

Figure 7. Frequency distribution of the gamma dose rate in Michika.

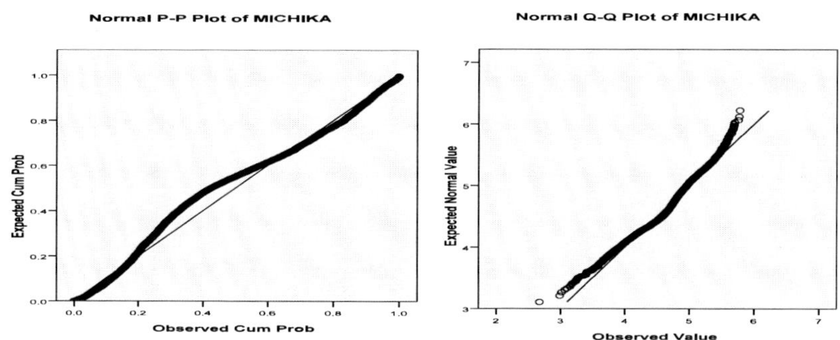

Figure 8. Log transformed data of normal P-P and Q-Q plots for Michika dose rate.

The high mean TGR dose rate value recorded in some areas (Figure 9) is due to the granitic geological features that are found in these areas. These areas are igneous acidic and extensively intruded by granitic rocks. The granite is relatively rich in radioactive minerals (UNSCEAR 2000).

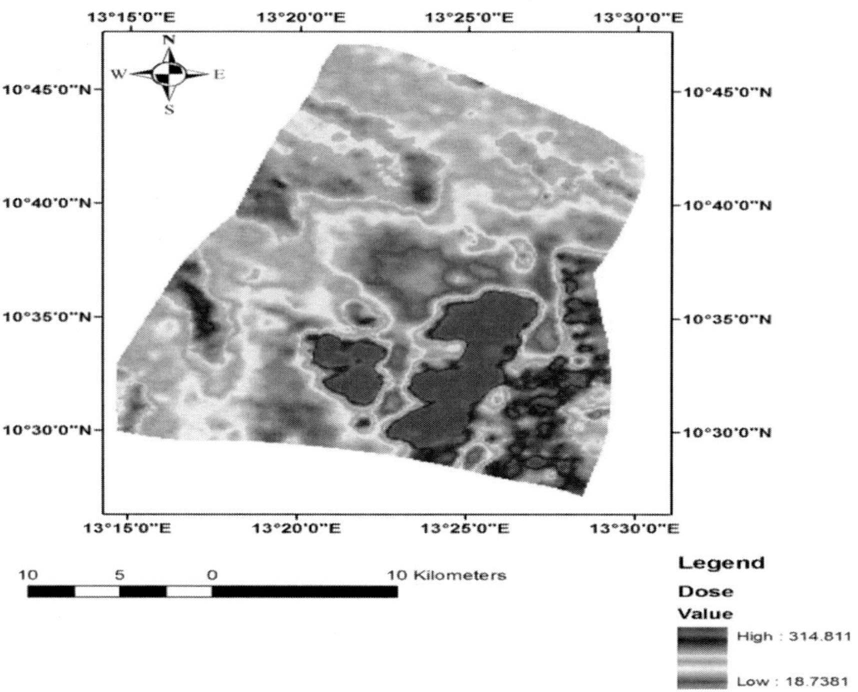

Figure 9. Isodose map of the study area.

Using the conversion coefficient factor for the absorbed dose in the air, we calculated an effective dose of 0.7 Sv Gy^{-1} and an outdoor occupancy factor of 0.2 (UNSCEAR 2000). The annual effective dose equivalent (AED) outdoor factor was calculated by using the following equation (1):

$$AED(mSv\, y^{-1}) = Dose\, rate(nGy\, h^{-1}) \times 24hours \times 365days \times OF \times 0.7 \times 10^{-6} - \quad (1)$$

where AED is the annual effective dose in mSv, and OF is the occupancy factor, which is 0.2 for outdoor environments. The average annual effective dose equivalent for the outdoors setting that was obtained in the study area was 0.141 mSv $y.^{-1}$, which is two times the world average annual effective dose value of 0.07 mSv y^{-1} (UNSCEAR 2000).

When considering the tissue weighting factor, the biological effect on the internal organs of a human body due to inhalation of radiation is calculated by using the following equation (2):

$$Effective\ dose(S = Absorbed\ Dose(Gy) \times tissue\ weigthing\ factor(WT\ ) \quad (2)$$

Table 3 shows the tissue or organ weighting factor for individual tissues or organs of the body. This factor provides variations in the effects of dose absorption by the specific tissue or organ of the body.

Table 3. Tissue weighting factors for individual tissues and organs (ICRP 2012)

Tissue or Organ	**Tissue weighting factor (W_T)**	$\sum W_T$
Gonads (testes or ovaries)	0.20	0.20
Red bone marrow, Colon, Lung, Stomach	0.12	0.48
Bladder, Breast, Liver, Oesophagus, Thyroid gland, Remainder**	0.05	0.30
Skin, Bone surfaces	0.01	0.02
TOTAL		1.00

** The remainder descriptor is composed of the following additional tissues and organs: adrenal gland, brain, upper large intestine, small intestine, kidney, muscle, pancreas, spleen, thymus and uterus.

Using equation (2), the effective doses that were received by each organ of the body from the mean absorbed dose rate of 115 nGy h^{-1} compared to the world average values with a mean dose rate of 59 nGy h^{-1} (UNSCEAR 2000) are given in Table 4.

Table 4. Effective doses received by each organ of the body in the study area compared to the one obtained by using the world mean absorbed dose rate

Tissue or Organ	W_T	Effective dose (μSvy^{-1}) (This study)	Effective dose (μSvy^{-1}) (world average)
Gonads (testes or ovaries)	0.20	0.023	0.012
Red bone marrow	0.12	0.014	0.007
Colon	0.12	0.014	0.007
Lung	0.12	0.014	0.007
Stomach	0.12	0.014	0.007
Bladder	0.05	0.006	0.003
Breast	0.05	0.006	0.003
Liver	0.05	0.006	0.003
Oesophagus,	0.05	0.006	0.003
Thyroid gland	0.05	0.006	0.003
Remainder**	0.05	0.006	0.003
Skin,	0.01	0.001	0.0006
Bone surfaces	0.01	0.001	0.0006
TOTAL	1.00	0.115	0.059

This shows that the individual organ effective dose in this study was two times greater than the world average values. The average lifetime effective dose (ALtED), cancer risk (R) and lifetime cancer risk (RL) per person were estimated by using the following equations (3, 4 and 5) (ICRP, 1990).

$$AL_tED\ (mSv) = AED(mSv) \times\ L_tE\ (y) \tag{3}$$

$$R = AED\ \times RF \tag{4}$$

$$R_L = L_tE \times R \tag{5}$$

where LtE is a life expectancy in years and RF is the risk factor (5.82 × 10-2 Sv-1) (BEIR VII 2006). The computed lifetime effective dose, cancer risk and lifetime cancer risk from outdoor exposure for each person living in the Michika area were 7.614 mSv, 8.21×10-3 and 4.43 x 10-1, respectively. These values are two times the world outdoor average values of 4.9 mSv, 4.07 × 10-3, and 2.85 × 10-1 for the previously mentioned respective factors. This considers a life expectancy of 54 years for Nigeria and 70 years for the world (WHO 2015).

Conclusion

The lifetime effective dose, cancer risk and lifetime cancer risk for persons living in Michika are twice the global average. The effective dose to each organ of the body due to inhalation of the gamma dose in the area was also found to be two times the global average. This computed effective dose and cancer risk values do not pose an immediate risk to individuals living in the area, but the continuous absorption of radiation over time may result in cancer-related ailments.

Acknowledgments

This project was funded by the tertiary education trust fund (TETFUND) of Nigeria under the 2014-2015 TETFUND Research Project Intervention for the 1st batch of Federal College of Education Yola.

References

BEIR VII, *The Biological Effects of Ionizing Radiation VII*. (2006). Health risks from exposure to low levels of ionizing radiation. Washington, D.C.: The National Academy of Sciences.

International Commission on Radiological Protection (ICRP). (1990). Recommendations of the International Commission on Radiological

Commission. *Annals of the ICRP*; 21(1-3). ICRP Publication 60. NY: Pergamon Press.

International Commission on Radiological Protection (ICRP). (2012). Compendium of Dose Coefficients based on ICRP Publication 60. ICRP Publication 119. *Ann. ICRP 41(Suppl.)* 2012 ICRP. Published by Elsevier Ltd.

James, J. W., Albert, Y. D. & Joel, M. K. (2015). Assessment of impact and adaptation strategies to climate change by local indigenous farmers of Michika local government area of Adamawa state, Nigeria. *European Scientific Journal* 11(26):1857 – 7881.

Nur, A. and Ayuni, K. N. (2011). Hydro-geophysical study of michika and environs, northeast Nigeria. *International Journal of the Physical Sciences*, 6(34): 7816 – 7827.

Stephen, W. R. & Misener, D. J. (2010). *Nigeria's nationwide high-resolution airborne geophysical surveys*. SEG Denver 2010 Annual Meeting.

United Nations Scientific Committee on the Effects of Atomic Radiation. (2000). *Report to the general assembly. Annex B: exposures from natural radiation sources.* (NY: UNSCEAR), ISBN-10: 9211422388 (2000).

World Health Organisation (WHO). (2015). *World heath statistic 2015*, http://www.who.int/countries/nga/en/.

In: The Influence of Ecosystem Services ... ISBN: 978-1-53619-977-2
Editors: Hasmah Abdullah et al.

Chapter 7

EFFECTS OF ACTIVE SMOKING AND SECOND-HAND SMOKE EXPOSURE AT HOME AMONG ADULTS IN THE RURAL AREA OF KAMPUNG BERIS LALANG, BACHOK, KELANTAN, MALAYSIA

Widad F.[1,2,*]***, Nur Fakhira Aida J.***[1]
and Hasmah Abdullah[1]
[1]School of Health Sciences, Health Campus Universiti Sains Malaysia, Kubang Kerian, Kelantan, Malaysia
[2]School of Industrial Technology, Universiti Sains Malaysia, Penang, Malaysia

ABSTRACT

Smoking is one of the most important public health issues among communities which may lead to health implications. Owing to this fact,

[*] Corresponding Author's E-mail: widad@usm.my.

this study was conducted to investigate the effects of smoking and secondhand smoke (SHS) exposure at home on the self-reported respiratory symptoms among adults in Kg. Beris Lalang, Bachok, Kelantan. A cross-sectional study among 218 respondents aged 18 - 65 years old was conducted using a purposive sampling through door to door survey. The socio-demographic variables examined were age, gender, occupation and level of education. The prevalence of smokers was associated with five respiratory symptoms including wheezing (46.8%), wheezing without cold (50.9%), chronic cough (63.2%), phlegm (36.7%) and sore throat (61.1%). Meanwhile, secondhand smoke exposure was associated with more prevalence in chest tightness (43.6%), coughing attack (43.2%), breathing trouble (46.7%) and sneezing, runny or blocked nose (40.5%) compared to non-smokers and non-exposed respondents (Chi-square test, $p < 0.001$). This study highlights the importance of minimising smoke exposure at home due to the prevalence of both active and passive smokers to develop respiratory symptoms related to tobacco smoke.

Keywords: active smoking, second-hand smoke, respiratory symptoms, rural area, Beris Lalang

INTRODUCTION

Smokers at home, as well as environmental tobacco smoke (ETS) or second-hand tobacco smoke (SHS), affect all household occupants because cigarette smoke remains in the air for the next 8 h after a person has finished the task of smoking (Juranić et al. 2018). Tobacco smoke contains various hazardous compounds, such as alcohols, phenols, ethers, hydrocarbons, sulfur organic compounds, lead, iron, copper, manganese, nickel, hydrocarbon molybdenum and other metabolites (Bird and Staines-Orozco 2016; Juranić et al. 2018), which are cumulatively detrimental to health. Previous studies have reported that ETS, also known as passive smoking, can cause respiratory illness in both adults and children (Larsson et al. 2003).

Respiratory problems, such as asthma, cough, chronic cough, phlegm, breathlessness, and wheezing, are significantly more likely to occur among smokers and those individuals exposed to SHS than their non-smoking counterparts (Bird and Staines-Orozco 2016). This study was conducted because an estimated 38.4% (7.6 million) of adults in Malaysia were exposed to second-hand smoke at home (IPH 2012). Despite several interventions being implemented, such as smoke-free areas, anti-smoking campaigns and smoking cessation programs in health clinics (Lim et al. 2013), the prevalence of smoking among male adults has increased; as well as an increasing trend of a higher prevalence of smoking in rural areas than in urban residences among Malaysians (Hum et al. 2016). In view of the impact of smoking as one of the most important public health issues among communities, this study was conducted to investigate the effects of smoking and SHS exposure at home on self-reported respiratory symptoms among adults in Kg. Beris Lalang, Bachok, Kelantan.

Materials and Methods

This study was conducted among 218 respondents from Kampung Beris Lalang, which is one of the residential areas located within mukim Telong, Bachok, Kelantan in the East Coast region of Malaysia. The sample size was calculated by using the single proportion formula, based on the prevalence study by Arya et al. (2012) and the value of the proportion by Khan et al. (2016). This cross-sectional study was performed on a sample of convenience from January to March 2019 and involved respondents aged between 18-65-years-old, which represents adults, based on the inclusion criteria of the study. Purposive sampling was performed through the use of self-administered questionnaires that were given through a door-to-door survey. The survey contained three sections: Section 1 contained questions pertaining to the residents' sociodemographic characteristics, Section 2 contained questions assessing the respondents' smoking practices and self-reported exposure to SHS; and Section 3 contained questions on respiratory symptoms.

The study was compared among three distinct groups: 1) non-smokers/non-exposed to SHS; 2) non-smokers/exposed to SHS; and 3) smokers.

This study adopted the validated questionnaire from St George's Respiratory Questionnaire by Jones (2009). The reliability of the questionnaire was evaluated by using the value of Cronbach's alpha, which was 0.800 for the current study, thus indicating an acceptable value between 0.70 and 0.95. Statistical analyses were conducted by using SPSS v24 for Windows. The investigators used chi-square tests ($\chi 2$) to determine the comparability between the three groups of respondents regarding the sociodemographic variables and smoking-related characteristics, which were included under the categorical data. Ethical approval was obtained from the Ethics Committee of Universiti Sains Malaysia (USM/JEPeM/18110638).

RESULTS AND DISCUSSION

Smoking Prevalence

Out of 218 respondents, smokers were least prevalent among the respondents from Kg. Beris Lalang (n = 32, 14.7%). This factor could be biased by the higher number of females (n = 141, 64.7%) than males (n = 77, 35.3%) who were located in the study area.

Out of these 32 smokers, most of the smokers were within the age range of 35 - 49-years-old (n = 10, 31.3%); among elderly individuals, most were within the age range of 50 - 65-years-old (n = 9, 28.1%). Elderly individuals (>60 years old) who smoked were found to have a prevalence of 39.2% in Malaysia and were associated with less knowledge and poor attitudes (Hum et al. 2016). Based on gender, 31% of the male respondents reported being smokers (n = 31), whereas only 1 female was reported as being a smoker (n = 1). The perception among young Malaysian rural women participants that smoking was harmful to the health of the person who smoked could have been the reason for the small number of female smokers.

Smoking prevalence varied by age, gender, occupation and education, with those individuals aged 35 - 49-years-old ($p < 0.001$) who were male respondents ($p < 0.001$), self-employed ($p < 0.001$) and having a tertiary education ($p < 0.001$) possessing a significantly higher smoking prevalence.

SHS Exposure

The overall SHS exposure of the respondents was 28% (n = 61). Approximately 34.5% of the respondents aged 35 - 49-years-old were exposed to SHS. Approximately 18% of the male respondents reported being exposed to SHS, compared to 82% of the females. Our study is similar with findings by Abdollahpour et al. (2019) showing that women are often more affected by exposure to SHS than men. Respondents aged 35 - 49-years-old ($p < 0.001$) and female respondents ($p < 0.001$) were significantly more likely to have been exposed to SHS than other age ranges and male counterparts (Table 1).

Respiratory Symptoms

The differences between the three distinct groups (i.e., non-smokers/non-exposed to SHS, non-smokers/exposed to SHS and smokers) were tested based on their self-reported frequencies of ten respiratory symptoms. Smokers were reported to have higher proportions of wheezing (46.8%), wheezing without cold (50.9%), chronic cough (63.2%), phlegm (36.7%) and sore throat within the past 12 months, compared to the other respiratory symptoms. Chest tightness, coughing attacks, breathing troubles, sneezing and runny or blocked noses were more prevalent among the non-smokers/exposed to SHS than among the non-smokers/non-exposed to SHS. Significant differences were identified with all of the respiratory symptoms (except breathing troubles and shortness of breath) when the non-smoker/non-exposed to SHS respondents were compared with those respondents who were non-smokers/exposed to SHS and those respondents who smoked (Table 1).

Table 1. Respiratory symptom variables by smoking status

Respiratory variables	**Smoking status**			**$\chi 2$ (p-value)**		
	Non-smoker/not exposed to SHS	Non-smoker/exposed to SHS	Smokers	A	B	C
Wheezing	7 (11.3%)	26 (41.9%)	29 (46.8%)	5.699	6.934	39.35
n=62				(0.017)	(0.008)	(<0.001)
Wheezing without cold	7 (12.3%)	21 (36.8%)	29 (38,5%)	4.655	8.265	34.438
n=57				0.031	0.004	<0.001
Chest tightness	7 (17.9%)	17 (43.6%)	15 (38.5%)	6.592	5.332	18.835
n=39				0.01	0.021	0.001
Coughing attack	15 (18.59%)	35 (43.2%)	31 (38.3%)	14.007	11.414	38.211
n=81				<0.001	<0.001	<0.001
Chronic cough	0	7 (36.8%)	12 (53.2%)	NA	NA	19
n=19						<0.001
Phlegm	0	5 (33.3%)	10 (66.6%)	NA	NA	15

Respiratory variables	Smoking status			χ2 (p-value)		
n=15						<0.001
Breathing trouble	5 (16.7%)	14 (46.7%)	11 (36.7%)	5.25	3.474	15.197
n=30				0.022	0.062	<0.001
Shortness of breath	4 (11.8%)	15(44.1%)	15(44.1%)	3.579	3.579	21.191
n=34				0.059	0.059	<0.001
Sore throat	0	14(38.9)	22(61.1%)	NA	NA	36
n=36						<0.001
Sneezing, runny or blocked nose	28 (37.8%)	30 (40.5%)	16 (21.6%)	30.711	12.426	13.918
n=74				<0.001	<0.001	<0.001

Notes: Respiratory symptoms are significantly different at the $P < 0.001$ level by using χ^2 tests.
A, test between non-smokers/non-exposed to SHS and non-smokers/exposed to SHS groups;
B, test between non-smokers/non-exposed to SHS and smokers groups;
C, test between non-smokers/exposed to SHS and smokers groups.
Abbreviation: SHS, second-hand smoke.

This finding indicates that active and passive smoking can be risk factors for the development of respiratory diseases, including lung cancer, chronic obstructive pulmonary disease, interstitial lung diseases and bronchial asthma (Ishii 2013).

Hence, people who smoke and who are regularly exposed to tobacco smoke may develop respiratory symptoms related to tobacco smoke.

Smoking status was directly related to the presence of respiratory symptoms (Amela et al. 2012). Multiple confounding factors may influence the presented respiratory symptoms, such as the effects of allergens, lifestyle and the history of one's health.

CONCLUSION

The current findings suggest that smokers had the highest prevalence of 5 respiratory symptoms, including wheezing, wheezing without cold, chronic cough, phlegm and sore throat. Passive smokers were associated with a higher prevalence of chest tightness, coughing attacks, breathing troubles, sneezing and runny or blocked noses than non-smokers and non-exposed respondents. This preliminary study provides initial baseline data about reported respiratory symptoms that are associated with exposures to active and passive smoking among adults within the home environment in a rural population. The least prevalent active and passive smokers within this population are a good sign for creating a smoke-free home preventing the initiation of tobacco use and eliminating exposure to SHS. This generic and basic study requires more in-depth assessments on the effect of smoking history, duration and types of tobacco use, effect of smoking among children and adolescents within the same household and comparisons of exposures at home and outside of the home to incorporate a more detailed understanding in the management of smoking and public health among local communities.

REFERENCES

Abdollahpour, I., Mansournia, M. A., Salimi, Y. and Nedjat, S. (2019). Lifetime prevalence and correlates of smoking behavior in Iranian adults'population; a cross-sectional study. *BMC Public Health*, 19(1056):1 - 11. https://doi.org/10.1186/s12889-019-7358-0.

Amela, D., Esad, P., Alen, D., Hasanbegovic, I., Mušanovic, J. and Spasojevic, N. (2012). The impact of respiratory symptoms presence on quality of life of tobacco smokers. *Arch. Pharma Pract., Official Publication of Archives of Global Professionals, Saudi Arabia*, 3(4):274 - 282.

Bird, Y. and Staines-Orozco, H. (2016). Pulmonary effects of active smoking and secondhand smoke exposure among adolescent students in Juárez, Mexico. *International Journal of Chronic Obstructive Pulmonary Disease*, 11: 1459 - 1467. doi:10.2147/COPD.S102999.

Hum, L. W., Hsien, C. C. M. and Nantha, Y. S. (2016). A review of smoking research in Malaysia. *Med. J. Malays*., 71: 29 - 41.

Ishii, Y. (2013). Smoking and respiratory diseases. Nihon Rinsho. *Japanese Journal of Clinical Medicine*, 71(3): 416 - 420.

Institute for Public Health (IPH). (2012). Report of the global adult tobacco survey (GATS) Malaysia, 2011, Ministry of Health Malaysia.

Juranić, B., Mikšić, Š., Rakošec, Ž. and Vuletić, S. (2018). Smoking habit and nicotine effects, smoking prevention and cessation, Mirjana Rajer, *IntechOpen,* DOI:10.5772/intechopen.77390. Availablefrom:https://www.intechopen.com/books/smoking-prevention-and-cessation/smoking-habit-and-nicotine-effects.

Jones, P. W., Harding, G., Berry, P., Wiklund, Chen W.-H. and Kline Leidy, N. (2009). Development and first validation of the COPD Assessment Test. *Eur. Respir. J.*, 34(3):648 - 654. https://doi.org/10.1183/09031936.00102509.

Larsson, H.-M., Loit, M., Meren, J., Põlluste, A., Magnusson, K. and Larsson, B. Lundbäck. (2003). Passive smoking and respiratory symptoms in the FinEsS Study. M.L. *European Respiratory Journal*, 21(4): 672 - 676. DOI: 10.1183/09031936.03.00033702.

Lim, H. K., Ghazali, S. M., Kee, C. C., Lim, K. K., Chan, Y. Y. and The, H. C. (2013). Epidemiology of smoking among Malaysian adult males: Prevalence and associated factors. *BMC Public Health*, 13(8): 1 - 10.

In: The Influence of Ecosystem Services … ISBN: 978-1-53619-977-2
Editors: Hasmah Abdullah et al.

Chapter 8

A PRELIMINARY STUDY OF DEPRESSION, ANXIETY, STRESS AND ITS FACTORS AMONG FIREFIGHTERS IN MALAYSIA

Ainul Husna K. [1,*], ***Khairilmizal S.*** [2]
and Najaatun Nisa A. R. [1]
[1]Institute of Medical Science Technology, UniKL, A1-1, Jalan TKS 1, Taman Kajang Sentral, Selangor, Kajang, Selangor, Malaysia
[2]School of Health Science, Health Campus Universiti Sains Malaysia, Kubang Kerian, Kelantan, Malaysia

ABSTRACT

Emergency responders have a higher risk of injury and death than other professions. Multiple factors should be identified because they are most vulnerable to depression, anxiety, and stress, which could have a negative impact on health. However, no such research has been conducted on Fire and Rescue Department Malaysia (FRDM) responders. Therefore, this study was conducted to assess depression, anxiety, and

* Corresponding Author's E-mail: ainulhusna@unikl.edu.my.

stress among FRDM responders, as well as their relationship to the factors that contribute to it. A questionnaire from DASS-21 and NIOSH-GJSQ was used to assess depression, anxiety, stress level, and its variables in 47 respondents from FRDM Kajang Station, Selangor. The finding showed that, 40.4% of respondents reported higher-than-average levels of depression, anxiety, and stress, while general health, social support from supervisors, and job satisfaction factors identified as contributing factors. In conclusion, social support from a co-worker as well as social support from family and friends may help respondents manage their depression, anxiety, and stress.

Keywords: emergency responders, depression, anxiety, stress

INTRODUCTION

Emergency responders are found to experience a higher level of stress than individuals employed in other sectors (Freeman et al. 2011). A study by Ainul Husna et al. concluded that there is an increasing number of challenges identified in a disaster exercise during a higher level of emergency response (Ainul Husna et al. 2016), thus indicating the presence of stress issues.

Each working sector has its own specific stressor (Smith et al. 2018; Wang et al. 2018). This situation includes emergency responders whose capabilities are determined by their ability to effectively manage the emergency (Khairilmizal et al. 2016c). Additionally, the factor of time plays an important role in responding to an emergency, as time will determine the outcome of the emergency response (Subramaniam et al. 2012). Hence, the ability to manage an emergency in a timely manner indicates the possibilities of emergency responders exposing themselves to stress and volatile environments (Deflem 2012). Eventually, some stressful events can result in emotional and psychological trauma being produced in responders (Roosli 2010).

Stress has been proven to exist during emergency or disaster responses (Paton and Flin 1999). Emergency responders who are mainly involved with emergencies or disasters are considered as having highly dangerous and stressful occupations (Skogstad et al. 2013; Smith et al. 2018). Hence, in reducing the exposure of emergency responders to stress issues, many tools have been developed to assist them in effectively managing emergencies (Khairilmizal et al. 2017a, 2017b).

The Fire and Rescue Department Malaysia (FRDM) is one of the main lead responding agencies that is entrusted with managing emergency situations in Malaysia (Khairilmizal et al. 2017b). In terms of expertise, equipment and capabilities, FRDM is one of the most recognized fire services in the world (Wahab 2011). However, the implementation of disaster management in a country is unique because each country has a unique governmental setup (Khairilmizal et al. 2016b; 2017c).

This uniqueness includes the FRDM, wherein the management of fire services in Malaysia is performed under the monitoring of the federal government, which eventually indicates challenges, such as the speediness of implementing new standards, allocations of firefighter workplaces and insuring all firefighters nationwide are at the same level of knowledge and expertise. Challenges also include financial support (Kusumasari et al. 2010; Salmon et al. 2011) because the FRDM is not directly under the federal government but is under the Ministry of Urban Wellbeing, Housing and Local Government. However, the centralization of FRDM under federal governments also suggests several management advantages, such as the centralization and standardization of trainings and equipment, as well as the implementation of standards nationwide.

To the researcher's discernments, stress issues and factors among emergency responders, especially among FRDM responders, have never been determined. Hence, it is the objective of this paper to provide preliminary studies in identifying depression, anxiety and stress among FRDM responders and its correlation with the factors contributing to it.

METHODOLOGY

The understanding of depression, anxiety, stress and its factors among FRDM personnel first requires systematic procedures in reviewing and evaluating the documents. For the purpose of this paper, the document analysis method (Khairilmizal et al. 2016e), face validity method (Hussin et al. 2012) and survey questionnaire method were used. Through document analysis, researchers can identify the meaning, gain understanding and develop empirical knowledge on depression, anxiety, stress and its factors.

A survey questionnaire was developed based on several predeveloped occupational stress questionnaires. The questionnaire was then validated by using face validity, which is defined as a test that appears to be valid or accepted by the researcher or field experts (Hussin et al. 2012). Four (4) disaster management and safety experts analysed, discussed and agreed that the developed questionnaire was relevant to be answered by FRDM personnel. However, there are possibilities that experts may take their knowledge for granted and may have different interpretations of the questionnaire. However, this can be ignored, as the questionnaire content is based on the following validated and adapted questionnaires:

- Depression, Anxiety and Stress Scale - 21 items (DASS-21) (Ronk et al. 2013)
- National Institute for Occupational Safety and Health Generic Job Stress Questionnaire (NIOSH-GJSQ) (Deguchi et al. 2016)

Including the demographic background, the questionnaire consisted of three (3) sections. In the other two (2) sections, the DASS-21 questionnaire was used to measure the emotional states of the depression, anxiety and stress of responders, whereas the NIOSH-GJSQ was used to measure the following risk factors (stressors):

- General health
- Workload
- Social support from supervisor
- Social support from co-worker
- Social support from family and friends
- Job satisfaction

The quantitative methods (survey) that we utilized involved 47 respondents. All of the respondents were personnel from the FRDM. They were all responders during emergencies from FRDM Kajang Station, Selangor. The respondents were chosen by using a purposive sampling technique, which is used when a limited number of people who have expertise in the area are being researched (Roosli 2010). Fire stations in the FRDM were categorized into three (3) grades, ranging from a low grade of "C" to a high grade of "A." FRDM Kajang Station is a grade "A" station, which indicates that the station has the capability and capacity to manage multiple types of incidents and to cover a highly vulnerable area, such as industrial areas, high rises and highly populated areas, as well as commercial areas. This provided advantages for this study, as firefighters working at these stations are exposed to a high number of daily responses, compared to grade "B" or grade "C" fire stations.

All 47 firefighters worked in shifts. For the purpose of standardizing the conditions of the survey questionnaire distribution, each respondent was required to answer the questionnaire at the start of their shifts, and only morning shift respondents were required to answer the questionnaire. Hence, the process of acquiring all of the data would take approximately one (1) week.

The data were descriptively analysed to determine the prevalence of occupational stress, and a Pearson correlation was used to determine the relationship between occupational stress risk factors and depression, anxiety and stress levels. The correlation could represent any value in the range [-1, 1]. The value can indicate no relationship ($r = 0$), a weak correlation ($0.1 < r < 0.3$), a moderate correlation ($0.3 < r < 0.5$) or a strong correlation ($r > 0.5$), with a significant value of $\rho < 0.05$ (Cohen, 1988).

RESULTS

A study of 47 FRDM respondents found that 40.4% of the FRDM responders had an above-normal level of depression, anxiety and stress. From the number of respondents that were previously indicated, they were categorized as having mild, moderate, severe and extremely severe levels of the disorders, based on the DASS-21 questionnaire.

Table 1. Levels of depression, anxiety and stress among the FRDM responders (N = 19)

Severity	Depression (%)	Anxiety (%)	Stress (%)
Mild	5.3	5.3	26.3
Moderate	10.5	0	21.1
Severe	21.1	15.8	26.3
Extremely Severe	63.2	78.9	26.3

From Table 1, it is observed that more than 80% of the respondents showed severe (21.1%) and extremely severe (63.2%) levels of depression. This indicates that there was a high level of depression among the FRDM responders. Furthermore, more than 90% of the respondents demonstrated a high level of anxiety, wherein the statistics demonstrated a 15.8% prevalence of severe anxiety and a 78.9% prevalence of extremely severe anxiety. Moreover, only 52.6% of the respondents showed severe and extremely severe stress levels, compared to both depression and anxiety. Researchers believe that stress is not a major issue among respondents, compared to depression and anxiety.

In terms of factors for depression, anxiety and stress, the previously mentioned descriptive statistical results were supported by the Pearson correlation test, as shown in Table 2. The table indicates that workload was not a factor, with 100% of the respondents showing no significant relationship ($r = 0$; $\rho < 0.05$). All of the respondents also expressed no relationship with depression, anxiety or stress from two other factors, including social support from coworkers ($r = 0$; $\rho < 0.05$) and social support from family and friends ($r = 0$; $\rho < 0.05$). Hence, it is in the opinion of the researchers that the workloads of all of the respondents is not a concern and that all of the respondents received a high degree of social support from coworkers, family and friends.

Eventually, general health, social support from supervisors and job satisfaction showed correlations with depression, anxiety and stress in the respondents, although the significance value was above 0.05 for the Pearson correlation test. The general health factor for stress level indicated a significant value ($\rho = 0.049$).

Table 2. Correlation and significance of depression, anxiety and stress towards its factors (N = 19)

		General Health	Workload	Social support from supervisor	Social support from co-worker	Social support from family and friends	Job Satisfaction
Depression	r	0.26	0	0.04	0	0	0.11
	ρ	0.29	0.001	0.86	0.001	0.001	0.65
Anxiety	r	0.07	0	-0.21	0	0	0.32
	ρ	0.78	0.001	0.39	0.001	0.001	0.19
Stress	r	0.46	0	0.33	0	0	0.11
	ρ	0.05	0.001	0.17	0.001	0.001	0.66

Table 3. Descriptive statistics of the factors that correlated with depression, anxiety and stress level (N = 19)

Factors	Low (%)	High (%)
General health	73.7	26.3
Social support from supervisor	15.8	84.2
Job satisfaction	5.3	94.7

Furthermore, Table 3 also indicates a weak correlation with depression for the factors of general health ($r = 0.26$; $\rho < 0.29$), social support from supervisors ($r = 0.04$; $\rho < 0.86$) and job satisfaction ($r = 0.11$; $\rho < 0.65$). For anxiety, the factors of general health and social support from supervisors indicated a weak correlation, but the factor of job satisfaction ($r = 0.32$) showed a moderate correlation. Moreover, for stress, the factors of general health ($r = 0.46$) and social support from supervisors ($r = 0.33$) indicated moderate correlations. Job satisfaction only indicated a weak correlation.

By screening the results from Table 35, the researchers identified that general health, social support from supervisors and job satisfaction have possible correlations with depression, anxiety and stress. Hence, the researchers conducted a descriptive statistical analysis of the factors that correlated with depression, anxiety and stress level. From Table 35, general health indicated that only 26.3% (SD = 0.45) of the respondents had high concerns for their health, which caused the identified correlation during the Pearson correlation test. For social support from supervisors, 15.8% (SD = 0.37) of the respondents stated that they received low support from their supervisors, thus indicating a factoring relationship with depression, anxiety and stress. Finally, 5.3% (SD = 0.23) of the respondents had low degrees of satisfaction with their job, which indicated that job satisfaction was one of the factors identified in this study.

DISCUSSION

When compared with the international framework on disaster management (Khairilmizal et al. 2016a), MNSC 20 has clearly stated that the FRDM is one of the lead responding agencies in Malaysia (Khairilmizal et al. 2016f). Although tools exist to assist responders in managing disasters (Khairilmizal et al. 2016d), FRDM officers and responders are exposed to various factors that lead to depression, anxiety and stress. The results, as shown in Table 35, indicate that the stress levels among responders showed a more linear distribution among mild, moderate, severe and extremely severe respondents. However, the results also indicate that stress has fewer issues than depression and anxiety.

The NIOSH-GJSQ questionnaire was adopted in this study to measure depression, anxiety and stress risk factors. The analysis, as shown in Table 36, indicated that workload factors did not contribute to depression, anxiety and stress in the respondents whatsoever. In fact, the researchers agree that social support from coworkers and social support from family and friends may contribute to assisting respondents in managing their depression, anxiety and stress. Eventually, general health, social support from supervisors and job satisfaction were identified as being contributing factors to depression, anxiety and stress in the respondents. However, the factors of general health, social support from supervisors and job satisfaction were specified only as weak contributing factors towards depression.

Although the analysis only indicated 2 weak and moderate correlations with depression, anxiety and stress, the Pearson correlation test results for general health, social support from supervisors and job satisfaction were supported by the descriptive statistics, as shown in Table 3. General health indicated that only a low percentage of respondents had a high degree of concerns about their health, thus causing the identified correlation during the Pearson correlation test. This is also the same in regards to social support from supervisors and job satisfaction, wherein a low percentage of the respondents stated that they received a low degree of support from the supervisor and were less satisfied with their job. This indicated the relationship of the factors towards depression, anxiety and stress, but only at low and moderate correlations.

In conclusion, this preliminary study of FRDM responders has proven that 40.4% of the responders had above-normal levels of depression, anxiety and stress. It is the researcher's opinion that while depression and anxiety among the respondents exhibited severe and extremely severe levels, stress showed a more linear distribution among mild, moderate, severe and extremely severe degrees, which indicates alarming concerns of depression and anxiety among the respondents.

It can also be concluded that general health, social support from supervisors and job satisfaction stressors have weak and moderate relationships with depression, anxiety and stress levels. However, only job satisfaction may contribute to major anxiety issues among the FRDM responders, whereas general health and social support from the supervisor may contribute to major stress issues among the FRDM responders. Eventually, general health stressors exhibited a significant correlation with stress levels.

The researchers agree that social support from coworkers and social support from family and friends may contribute to assisting respondents in managing their depression, anxiety and stress. However, a further detailed study needs to be conducted. Although this preliminary study did indicate certain significant results, researchers agree that a larger sample size will contribute to more conclusive results.

This study was a part of disaster management research in Malaysia, and the results indicate that there are significant depression, anxiety and stress levels among firefighters. However, the acquired data only indicate a preliminary sample from a grade "A" fire station. As levels of depression, anxiety and stress will affect the overall capabilities of firefighters in effectively managing disasters (Khairilmizal et al. 2016b), and this study indicates that a further detailed study in determining the levels of depression, anxiety and stress among firefighters nationwide is feasible.

ACKNOWLEDGMENTS

We would like to thank the Universiti Kuala Lumpur (UniKL), Institute of Medical Science Technology, for support in completing this study. We also wish to thank the Fire and Rescue Department Malaysia and Universiti Sains Malaysia under the Disaster Management Research collaboration for providing the opportunity for doing the previously described study.

CONFLICT OF INTEREST

No potential conflict of interest relevant to this article was reported.

REFERENCES

Ainul Husna, K., Siti Hawa, B., Hussin, M. F. & Khairilmizal, S. (2016). Effective Emergency Management: A Malaysian public listed oil and gas company perspective, In: *International Medical Science*

Technology Conference (IMSTC2016). UniKL MESTECH, Kuala Lumpur.

Cohen, J. (1988). Statistical power analysis for the behavioral sciences. *Stat. Power Anal. Behav. Sci.* https://doi.org/10.1234/12345678.

Deflem, M. (2012). Introduction: Disasters and hazards in socio-legal studies, In: *Sociology of Crime Law and Deviance*. Emerald Group Publishing Ltd., pp. ix-xii. https://doi.org/10.1108/S1521-6136(2012) 0000017003.

Deguchi, Y., Iwasaki, S., Konishi, A., Ishimoto, H., Ogawa, K., Fukuda, Y., Nitta, T. & Inoue, K. (2016). The usefulness of assessing and identifying workers' temperaments and their effects on occupational stress in the workplace. *PLoS One*, 11: 1-12. https://doi.org/10. 1371/journal.pone.0156339.

Freeman, J., Vidgen, A. & Davies-edwards, E. (2011). Staff experiences of working in crisis resolution and home treatment. *Rev. Lit. Arts Am.*, 16:76-87. https://doi.org/10.1108/ 13619321111158016.

Hussin, M. F., Wang, B. & Hipnie, R. (2012). The reliability and validity of basic offshore safety and emergency training knowledge test. *J. King Saud. Univ. - Eng. Sci.,* 24: 95-105. https://doi.org/10.1016/j. jksues.2011.05.002.

Khairilmizal, S., Hussin, M. F., Ainul Husna, K., Hussain, A. R., Jusoh, M. H. & Sulaiman, A. A. (2016a). Implementation of disaster management policy in Malaysia and its compliance towards international disaster management framework. *Information*, 19: 3307-3311.

Khairilmizal, S., Hussin, M. F., Ainul Husna, K., Yassin, A. I. M., Jusoh, M. H., Sulaiman, A. A., Wan Ahmad Syafiq, W. A. H. & Saadun, J. (2016b). The need for an integrated disaster management system for lead responding agency in Malaysia during response and recovery phase. *Information*, 19: 3307-3311.

Khairilmizal, S., Hussin, M. F., Ainul Husna, K., Yassin, A. I. M., Wan Ahmad Syafiq, W. A. H., Jusoh, M. H. H., Sulaiman, A. A. & Saadun, J. (2016c). The need for an integrated disaster management system for lead responding agency in Malaysia during response and recovery phase. *Information*, 19: 3307-3311.

Khairilmizal, S., Hussin, M. F., Ainul Husna, K., Yassin, A. I. M., Wan Ahmad Syafiq, W. A. H., Saadun, J. & Hussain, A. R. (2016d). Model of an incident potential index and integrated disaster management online system for lead responding agency in Malaysia during response and recovery phase, In: Arshad, A. S., Hai, A. H. A., Samsudin, M. R. (Eds.), *Technology and Applications for Disaster Management 2nd International Conference*. Astronautic Technology (M) SDN. BHD., Shah Alam.

Khairilmizal, S., Hussin, M. F., Yassin, A. I. M., Ainul Husna, K., Sulaiman, A. A., Jusoh, M. H. & Mohd Haikal, K. (2016e). Evolution of disaster and disaster management policy in Malaysia. *Adv. Sci. Lett.*, 22. https://doi.org/doi.org/10.1166/asl.2016.8107.

Khairilmizal, S., Hussin, M. F., Yassin, A. I. M., Hussain, A. R., Ainul Husna, K., Jusoh, M. H., Sulaiman, A. A., Saadun, J. & Mohd Haikal, K. (2016f). Policy on disaster management in Malaysia : The Need of Supporting Governance. *Adv. Sci. Lett.*, 22. https://doi.org/10.1166/asl.2016.8108.

Khairilmizal, S., Hussin, M. F., Ainul Husna, K., Yassin, A. I. M., Jusoh, M. H., Sulaiman, A. A., Wan Ahmad Syafiq, W. A. H. & Saadun, J. (2017a). Criteria for an Integrated disaster management system for lead responding agency in Malaysia. *Adv. Sci. Lett.*, 23: 4278-4280. https://doi.org/https://doi.org/10.1166/asl.2017.8248.

Khairilmizal, S., Hussin, M. F., Yassin, A. I. M., Ainul Husna, K., Jusoh, M. H., Sulaiman, A. A., Saadun, J., Mohd Haikal, K. & Hussain, A. R. (2017b). Design and development of strategic disaster management database for lead responding agency in Malaysia during response and early recovery phases. *Int. J. Electr. Electron. Syst. Res.*, 10.

Khairilmizal, S., Hussin, M. F., Yassin, A. I. M., Salleh, M. K. M., Ainul Husna, K., Jusoh, M. H., Sulaiman, A. A., Saadun, J., Mohd Haikal, K. & Hussain, A. R. (2017c). Incident potential index model complimenting Malaysia disaster management environment, In:

International Conference on Engineering Management (ICEM 2017). IPN, Kuala Lumpur.

Kusumasari, B., Alam, Q. & Siddiqui, K. (2010). Resource capability for local government in managing disaster. *Disaster Prev. Manag.* 19: 438-451.

Paton, D. & Flin, R. (1999). Disaster stress: an emergency management perspective. *Disaster Prev. Manag.* 8: 261-267.

Ronk, F. R., Korman, J. R., Hooke, G. R., & Page, A. C. (2013). Assessing clinical significance of treatment outcomes using the DASS-21. *Psychol. Assess.* 25, 1103-10. https://doi.org/10.1037/a 0033100.

Roosli, R. (2010). *Managing disasters in Malaysia: the attitude of officials towards compliance with the MNSC Directive 20.* University of Northumbria, Newcastle.

Salmon, P., Stanton, N., Jenkins, D. & Walker, G. (2011). Coordination during multi-agency emergency response: issues and solutions. *Disaster Prev. Manag.* 20: 140-158.

Skogstad, M., Skorstad, M., Lie, A., Conradi, H. S., Heir, T. & Weisth, L. (2013). Work-related post-traumatic stress disorder. *Occup. Med.* (Chic. Ill). 63: 175-182. https://doi.org/10.1093/occmed/kqt003.

Smith, T. D., Hughes, K., DeJoy, D. M., & Dyal, M. A. (2018). Assessment of relationships between work stress, work-family conflict, burnout and firefighter safety behavior outcomes. *Saf. Sci.* 103: 287-292. https://doi.org/10.1016/j.ssci.2017.12.005.

Subramaniam, C., Ali, H. & Shamsudin, F. M. (2012). Initial emergency response performance of fire fighters in Malaysia. *Public Sect. Manag.* 25, 64-73. https://doi.org/10.1108/09513551211200294.

Wahab, M. T. A. (2011). *Malaysia National progress report on the implementation of the Hyogo*, UNISDR Publication. Geneva.

Wang, D., Wang, X. & Xia, N. (2018). How safety-related stress affects workers' safety behavior: The moderating role of psychological capital. *Saf. Sci.* 103: 247-259. https://doi.org/10.1016/j.ssci.2017.11.020.

In: The Influence of Ecosystem Services ... ISBN: 978-1-53619-977-2
Editors: Hasmah Abdullah et al.

Chapter 9

COMPOSTING USING DIFFERENT TYPES OF ORGANIC WASTE FROM SELECTED CAFETERIAS IN UNIVERSITI SAINS MALAYSIA HEALTH CAMPUS AND ITS EFFECT ON *IPOMOEA AQUATICA* GROWTH

Raja Nur Shafieza R. M., Nurulilyana S.*[*], *Widad F.*, *Nur Fatihah R. and Nurkhairina Syahirah S.
School of Health Science, Health Campus Universiti Sains Malaysia,
Kubang Kerian, Kelantan, Malaysia

ABSTRACT

Composting is the natural recycling of organic matter. The process takes several months to produce a high-quality mature compost that can be used as a fertilizer. This study was designed to track the time it takes for different types of compost to mature and to compare the growth of *Ipomoea aquatica* after 21 days of planting with different types of

[*] Corresponding Author's E-mail: nurulilyana@usm.my.

compost. The levels of total nitrogen, total phosphorus, and potassium in the various types of compost were also compared. Organic waste was collected from the cafeteria at the Universiti Sains Malaysia Health Campus. Fruit peels, food leftovers, and mixed organic waste were allowed to decompose for 60 to 90 days. The mature compost was then used as a planting medium for *Ipomoea aquatica*. Acid digestion was used for macro element analysis. The potassium level was determined using atomic absorption spectroscopy (AAS), while total nitrogen and total phosphorus were determined using a COD Reactor DRB200 and a colorimeter (Hach Model DR/890). The results showed that the time required to mature compost from fruit peels was 60 days, while the others took 75 days. Furthermore, there was a significant difference in *Ipomoea aquatica* growth when different types of organic waste compost were used. There was also a significant difference in macro elements between the various types of organic waste compost. These findings indicate that composts made from various organic wastes differed in terms of quality and stability, which is further influenced by the composition of the raw material used in compost production.

Keywords: Compost, Organic waste, Recycle, Environment, *Ipomoea aquatica*

INTRODUCTION

Organic waste contains materials originating from living organisms. Such waste is often disposed with other wastes in landfills or incinerators, although since they are biodegradable, some organic wastes are suitable for composting and land application (Thomson Gale 2003). Organic waste can be classified into many types and is generated by food processing, agriculture, municipalities and industry. Organic waste commonly includes food waste, fruit and vegetable peels and flower trimmings. Some of the organic materials in municipal solid waste are separated before disposal for purposes other than composting. For instance, paper and cardboard are commonly removed for recycling. However, without proper management of such organic waste, a number of environmental problems could occur. The high amount of food waste is the main cause of most issues related to landfills, such as foul odour, toxic leachates, greenhouse gas emissions and vermin infestation (Kadir et al., 2016). In Malaysia, such issues are a major concern because research has shown that waste by the Malaysian population is sufficient to feed millions daily. Every year, an average Malaysian household throws away more than one month's salary on food that they do not eat (Solid Waste Corporation Management (SWCorp) 2017). This waste is equivalent to RM 2,700 a year, which is more than the mean monthly salary for an individual in an urban area of RM 2,400. Another study conducted by SWCorp also showed that Malaysians generated 38,000 tonnes of solid waste daily in 2016, of which 15,000 tonnes was food waste (The Star Online 2016).

This problem could be addressed by performing composting, which is the best low-cost alternative solution for municipal waste because the composting method can degrade many types of organic wastes, such as fruit peels, vegetables, plants, food waste and yard wastes. The composition of organic waste provides many benefits because it can be used as a nutrient source for crops, as a soil additive and for environmental management.

However, many factors could affect the quality of composted products because different types of organic wastes might have different concentrations of nutrients, nitrogen (N), phosphorus (P) and potassium (K), which are the common macronutrients present in fertilizers (Kadir et al., 2016). In terms of the factors affecting the composting process, the temperature, pH, moisture contents and carbon nitrogen ratio (C:N) are the main parameters that could contribute to the efficiency of the composting process (Fathi 2014).

This study was conducted at the Universiti Sains Malaysia Health Campus, where organic waste was collected from selected cafeterias in this area. This study can encourage cafeteria owners to focus on preparing their organic waste. The operators needed to provide permission and cooperate with the authors of this study for the collection of organic waste from their cafeteria. Their support and organic waste contribution are very important. To obtain their permission, this study will provide them with an explanation of the study objectives. The importance of the study's objectives can help cafeteria owners realize that they are responsible for improving their knowledge, action and practice on the use of organic waste. They need to know how organic waste will affect the environment if it is not managed properly.

In addition, it is important to perform composting because this practice provides many benefits to the environment. Most importantly, composting is beneficial to the environment because organic waste disposed in landfills will emit methane gas and carbon dioxide, which trap heat in the atmosphere. Composting can also improve the soil contents and nutrients required for harvesting plants, which can reduce the use of synthetic fertilizers. In addition, the significance of this study is to promote recycling activity. Composting is a part of recycling and referred to as a natural method of recycling.

METHODOLOGY

Study Design

This study implemented a cross-sectional study design and was conducted at the Environmental Occupational Safety & Health (EOSH) Centre, Universiti Sains Malaysia Health Campus. Organic waste was collected at the cafeteria, and then organic waste segregating and composting procedures were performed. Organic waste was collected from September 2018 until June 2019, and the first composting process started on 30 December 2019, with the composting period lasting for approximately 60 to 75 days (Aeslina et al., 2016). The mature compost was ready to use in March 2019, and cultivation of Ipomoea aquatica was then started, which was followed by an analysis of the nutrient content.

Material and Methods

This study was divided into two stages. In the first stage, organic wastes collected from cafeterias were segregated into several types. Then, the composting process started. After 60 to 75 days, the compost had already matured and was then submitted for macroelement analysis. The results were compared with standard readings on the available organic compost. In the second stage, the end products of compost obtained from each type of organic waste were used as the medium for planting Ipomoea aquatica. It was also planted using common organic compost to compare its effectiveness.

First Stage Description

Three compost bins were set up in a courtyard of the EOSH Centre. The three different types of organic wastes are fruit peels, food leftovers and mixed. The same level of each organic waste at a depth of 5 cm was put into a special bin that was labelled. First, the base of that bin was covered with woodchips at an approximate depth of 5 cm. The next layer was green leaves and brown dry leaves, followed by organic waste and dry leaves again. These layers alternated with available organic compost as the medium to facilitate the composting process. The steps were repeated to build a heap layer upon layer. It was then covered with organic compost to make the process of composting faster. The last surface was covered with dry leaves and then moistened slightly. The bins were then closed and left for 60 to 90 days. Every 14 days, the heap layer was mixed to ensure that the organic wastes were well mixed during the composting process.

After 60 to 90 days, the composts were completed and matured. An obvious indicator of mature compost is when the colour darkens and releases odour from the soil. It is then used for a macro element analysis. First, the compost undergoes acid digestion. Next, potassium was analysed by atomic absorption spectroscopy (AAS), while total nitrogen and total phosphorus were analysed by a COD Reactor DRB200 and colorimeter (Hach Model DR/890).

Second Stage Description

The final product of the compost process was used as a medium for planting water spinach, which is known as Ipomoea aquatica. This plant was watered twice a day, in the morning and in the evening. The seeds used were of the same type to ensure a consistent quality of the plants.

The growth of this plant was monitored after 21 days of planting based on the different types of organic waste compost and commercially available organic compost as the reference. Plant growth was monitored based on the length of the plant.

Statistical Analysis

The collected data were analysed using Statistical Package for Social Sciences (SPSS) version 22.0. Descriptive statistics were applied to assess the time required for the compost to mature. One-way ANOVA was used to evaluate the differences in the length of Ipomoea aquatica cultivated with different types of organic waste compost. The Kruskal-Wallis test is a nonparametric test for more than two independent samples, and it was used to compare the macro elements in the compost, including the total nitrogen, total phosphorus and potassium contents, according to the different types of composts.

RESULTS

Time Required for Compost to Mature

Observation was performed daily. Differences between the composting processes occurred every 15 days for fruit peel compost collected from the cafe. On the 60th day, the texture of the compost was identified, and it had soil-based textures and characteristics. For the leftover-type compost and mixed-type compost, the composting process slowly occurred in the container. The time required to complete the composting process differed based on the fruit peel type. On the 75th day, the texture of the compost was identified, which was slightly moist.

Ipomoea Aquatica Growth

Ipomoea aquatica growth was measured based on the length of the plant after 21 days of planting. Table 1 shows the results for the length of Ipomoea aquatica.

Table 1. Growth of Ipomoea aquatica

Cafe	Types of compost	Length of *Ipomoea aquatica* (cm)		
		Reading 1	Reading 2	Reading 3
Murni	Fruit peels	17.80	17.90	17.90
	Leftovers	16.60	16.70	16.70
	Mixed	18.30	18.20	18.00
Nurani	Fruit peels	17.90	17.70	17.60
	Leftovers	16.50	16.60	16.50
	Mixed	18.50	18.40	18.40
Harmoni	Fruit peels	17.40	17.30	17.30
	Leftovers	16.30	16.20	16.10
	Mixed	18.60	18.50	18.70
	Available organic compost	20.10	19.90	20.00

Comparison of the Growth of Ipomoea Aquatica Cultivated Using Different Types of Organic Waste Compost

Differences in Ipomoea aquatica growth cultivated using different types of organic waste compost are shown in Table 2. The p-value was less than 0.05, which indicated significant differences in the mean length of Ipomoea aquatica. In addition, the Scheffe post hoc test was performed to determine the differences for each pair, and it was applied because the test of homogeneity variances and Levene's test indicated that the p-value was greater than 0.05. The post hoc test results showed that all pairs were significantly different at p-value < 0.001.

Table 2. Statistical data from the one-way ANOVA

Type of compost	Mean (SD) length of *Ipomoea aquatic*	F statistics (df)	p-value*
Fruit peels	17.64 (0.26)	229.218(3)	< 0.001
Leftovers	16.47 (0.22)		
Mixed	18.40 (0.21)		
Available organic compost	20.00 (0.10)		

* Significant difference at $p < 0.05$, statistical test – one-way ANOVA.

Table 3. Statistical data for the Kruskal-Wallis test

Variable	Median (IQR)			Chi-square statistic (df)	*p*-value
	Fruit peels	Leftovers	Mixed		
Total nitrogen	28.80 (0.10)	27.50 (0.05)	28.60 (0.15)	24.369 (3)	< 0.001*
Total phosphorus	98.40 (17.50)	110.00 (24.15)	110.00 (24.95)	4.984 (3)	0.173
Potassium	2.55 (0.01)	2.34 (0.05)	2.50 (0.02)	24.200 (3)	< 0.001*

* Significant difference at $p < 0.001$, statistical test – Kruskal-Wallis test.

Comparison of Total Nitrogen, Total Phosphorus and Potassium in Three Different Types of Compost

The Kruskal-Wallis test is a nonparametric test for more than two independent samples. This test was used to compare the macro elements in compost, including total nitrogen, total phosphorus and potassium, according to the different types of composts. The p-value was less than 0.05 for total nitrogen and potassium. Thus, there was a significant difference in the median contents of these macro elements in the compost, as shown in Table 3. However, the total phosphorus results demonstrated a p-value of 0.173 ($p > 0.05$) (Table 3).

Discussion

Mature compost is a dry material with a dark colour and a relatively small particle size, and it has an odour similar to that of soil (Oviedo-Ocana et al., 2017). These characteristics were obtained by typical on-site quality criteria. Based on a previous study, Kadir et al., (2016) stated that compost would take 3 and 6 months to mature and could then be used as organic fertilizer. This finding was inconsistent with the results of this study, which was because they used tapioca peels as the substitute for fermented soybeans to produce fermentation liquid. In addition, they also used grated coconut as the fermentation bed. Grated coconuts were used as a substitute for rice husks as a nitrogen source and soil as a carbon source in the fermented bed. However, in this study, the fruit peel types were mostly banana peels, watermelon peels, orange peels, papaya peels and apple peels. Moreover, fermentation liquid was not added to the compost and other materials were not added that could affect the process.

For the leftovers and mixed-type compost, both needed more than 60 days to completely decompose and 75 days or more to form a mature compost. The time taken for this process was longer than that for fruit peel-type compost because of the organic waste composition in the container. Leftovers put in the container contained processed food waste with oily and moist material. These factors cause the process to take more than 60 days to decompose. According to Oviedo et al., (2017), the physicochemical characteristics of unprocessed food waste and processed food waste are not similar. For instance, the microbial population is expected to be different between cooked and uncooked food. This feature can influence the composting process and product quality.

In addition, according to Azim et al., (2018), the proportions of carbon and nitrogen in composting material are important. Carbon, serves as both a source of energy and an elemental component for microorganisms and nitrogen and is essential for the synthesis of amino acids, proteins and nucleic acids. Due to the recalcitrant carbon, which is difficult to degrade, the compost of shrubs and wood in that study took a longer time to mature, which was 18 months compared to a compost of household waste that matured in 7 months. Thus, in this study, the amount of organic waste, dried leaves, and ready compost that were placed layer by layer were the same for all types of compost, and the same size and type of container were also used. Each alternating layer of organic waste and dried leaves heap was 5 cm. As predicted, the leftover-type compost and mixed-type compost needed more than 60 days to completely decompose, which was longer compared to the fruit peel-type compost because the fruit peel-type compost only consisted of raw organic waste and did not include processed food waste.

Thus, to ensure an increase in plant or vegetable growth, it was important to promote a higher bioaccumulation of macro- and micronutrients from the soil compared to commercial chemical fertilizer. Although the composts produced in this study were not as effective as the commercially available compost, they could still aid in supplying nutrients to plants.

In this study, the nitrogen content of composts varied according to the source material and how it was composted. In general, nitrogen becomes less available as the compost matures with nitrogen-rich feedstock but more available with carbonaceous feedstock (Mangan et al., 2013). Therefore, the leftovers and mixed compost also contained the same amount of organic waste and dried leaves during the composting process, which affected the total nitrogen reading of all samples. The composition was the same for all the samples during the composting layer process.

Phosphorus is lacking in most composting materials, which makes compost a poor substitute for a balanced fertilizer in gardens. The most notable exception is manure, which is typically high in phosphorus and potassium as well as nitrogen. However, manure was not used in this study. Manure is one of the most balanced compost materials available and can even be used as an organic fertilizer for yards (Douglas 2018).

Potassium in finished compost is much more available for plant uptake than nitrogen and phosphorus since potassium is not incorporated into organic matter. However, much of the potassium can be leached from the compost since it is water soluble (Mangan et al., 2013).

CONCLUSION

First, this study was able to monitor the time required to achieve composting end product characteristics based on different types of organic waste. According to the findings, the fruit peel-type compost only took 60 days to completely decompose and become mature compost, which is observed when the colour changes to dark and the odour is consistent with soil. Additionally, the leftovers and mixed-type compost both required 75 days to completely decompose. This study also compared differences in the growth of *Ipomoea aquatica* cultivated using different types of organic waste compost. The findings showed that there was a significant difference between *Ipomoea aquatica* planted using different types of organic waste compost. It can be concluded that the plants grew efficiently using the available organic compost, followed by the mixed-type compost, fruit peel compost and leftover-type compost. This study was also performed to compare the macroelements total nitrogen, total phosphorus and potassium in the different types of organic waste compost. The results showed that there was a significant difference in the total nitrogen and potassium of the different types of organic waste compost while no significant difference was observed in the total phosphorus. Thus, composts prepared from different organic wastes differed in their quality and stability, which further depended upon the composition of the raw material used for compost production.

Acknowledgments

The authors would like to acknowledge Universiti Sains Malaysia for supporting this study. In addition, we would like to acknowledge all the health personnel, professionals and individuals involved in this study.

References

Acero, L. H., (2013). Growth response of *Brassica rapa* on the different wavelength of light. *International Journal of Chemical Engineering and Applications*, 4(6): 415–418.

Aeslina, A. K., Nur, W. A. & Siti, N. J. (2016). An overview of organic waste in composting. *MATEC Web of Conferences*, 47:1-6.

Alvarado, D., Buttrago, E., Sole, M. & Frotado, K. (2008). Experimental evaluation of a composted seaweed extract as microalgal culture media. *Aquaculture International,* 16:85.

Ana, K. d. C. F., Nildo, d. S. D., Francisco, S. d. S. J., Daianni, A. d. C. F., Cleyton, d. S. F. & Tiago, d. S. L. (2017), Composting of household organic waste and its effects on growth and mineral composition of cherry tomato, An Interdisciplinary *Journal of Applied Science*, 47: 2-6.

Azim, K., Soudi, B., Boukhari, S., Perissol, C., Roussos, S. & Thami Alami. I. (2018). Composting parameters and compost quality: a literature review. *Organic Agriculture*, 8(2):141–158.

Enriquez, V. A. (2017). Growth performance of water spinach, *Ipomoea aquatica* on seaweed, *Eucheuma cottonii* compost treated soil and other commercial growing media. *International Journal of Agriculture and Economic Development*, 5(1): 29-37.

Fathi, H. (2014). Municipal solid waste characterization and it is assessment for potential compost production: A case study in Zanjan City, Iran. *American Journal of Agriculture and Forestry*, 2(2): 39.

Haq, T., Khan, F. A., Begum, R. & Munshi, A. B. (2011). Bioconversion of drifted seaweed biomass into organic compost collected From the Karachi Coast Pak. *Journal of Botany*, 43(6): 3049.

Hongfei, F., Xie, B. & Pan, S. (2011). Evaluation of antioxidant activities of principal carotenoids available in water spinach (*Ipomoea aquatica*). *Journal of Food Composition and Analysis*, 24(2): 288-297.

Jamaludin, S. N., Abdul Kadir, A. & Azhari, N. W. (2017). Study on NPK performance in food waste composting by using agricultural fermentation *in* Zainorizuan, M. J. et al., (eds.). *MATEC Web of Conferences*, 103:05015.

Kadir, A. A., Azhari, N. W. & Jamaludin, S. N. (2016). *An overview of organic waste in composting*. In Abd Rahman, N., Mohd Jaini, Z., Yunus, R. & Rahmat, S. N., (eds.). *MATEC Web of Conferences*, 47:05025.

Kibria, G. (2017). *Food waste impacts on climate change and water resources.* A part of a research project on community-based environmental and sustainability education model in Australia, 1-4.

Mangan, F., Barker, A., Bodine, S. & Borten, P. (2013). *Compost use and soil fertility.* University of Massachusetts Extension Amherst Publication Agriculture and Landscape Program, p.1-4.

Md, A. I., Golam, F., Ayasha, A., Md, M. H. & Dilip, N. (2017). Effect of organic, inorganic fertilisers and plant spacing on the growth and yield of cabbage. *Agriculture*, 7(31):2-6.

Oviedo-Ocaña, E. R., Dominguez, I., Komilis, D. & Sánchez, A., (2017). Co-composting of green waste mixed with unprocessed and processed food waste: influence on the composting process and product quality. *Waste and Biomass Valorization*, 10(1): 63–74.

The Star Online (2016). *Research shows Malaysians waste enough to feed millions daily -Nation* [Online]. Retrieved on September 7, 2018 from https://www.thestar.com.my/news/nation/2016/05/31/food-and-money-down-the-drain-research-shows-malaysians-waste-enough-to-feed-millions-daily/.

Thomson, G. (2003). *Environmental Encyclopedia*, Bukupedia, United States of America, p.927.

Vegetables and Cash Crops Statistic Malaysia, (2018). Report of the hectarage and production of vegetables, Malaysia by Types, 2017 – 2018, p.65.

Winberg, P. C., Demestre, C. & Willis, S. (2011). Evaluating microdictyon umbilicatum bloom biomass as an agricultural compost conditioner for native and commercial plants. *Report to Shoalhaven City Council*, 30p.

Zeman, C., Depken, D. & Rich, M. (2002), Research on how the composting process impacts greenhouse gas emissions and global warming. *Compost Science & Utilization Journal*, 10(1):72-86.

In: The Influence of Ecosystem Services ... ISBN: 978-1-53619-977-2
Editors: Hasmah Abdullah et al.

Chapter 10

IN SILICO MODELLING OF THE CORE CATALYTIC SITE OF CHITIN DEACETYLASE FROM *ASPERGILLUS LUCHUENSIS*

Nor Amy Haryatin A.[1]***, Ragheed Hussam Y.***[1]***,***
Fatahiya M. T.[2] ***and Nurul Bahiyah A. K.***[1,*]
[1]Malaysia Japan International Institute of Technology,
Universiti Teknologi Malaysia, Malaysia
[2]Faculty of Chemical Engineering, Universiti Teknologi Mara
Terengganu Branch, Bukit Besi Campus, Terengganu, Malaysia

ABSTRACT

This study predicts the three-dimensional structure of chitin deacetylase using the method of homology modeling. The objective of this study is to compare the predicted structure with the native experimentally solved structure. This method involves the identification of template protein, alignment of the target sequence to the template, generation of the 3D model, and model evaluation. In addition to that, the location of the catalytic sites was predicted as well. The accuracy of this

[*] Corresponding Author's E-mail: r-bahiah@utm.my.

method was highly dependent on the percentage of similarity between the target and the template protein. In conclusion, the computational method complements the experimental structure in predicting protein structures.

Keywords: homology modelling, catalytic site, chitin deacetylase, structure prediction

Introduction

A large number of new protein sequences have been found in humans, plants and other organisms, and they are known as amino acid sequences or primary sequences. To determine the function of each protein, the three-dimensional structure needs to be predicted. The aim of structural genomics is to delineate or predict the three-dimensional structure of a protein encoded by the genome. Although exploratory strategies, such as X-ray crystallography and nuclear magnetic resonance (NMR) spectroscopy, are accurate for predicting the protein structure, they are problematic because of the high time and cost requirements (Tramontane 1998).

In the previous 50 years, tremendous advancements have been achieved in protein three-dimensional structure modelling; however, such work has not kept pace with the development of protein sequence data (Marks et al., 2012). This discrepancy will result in large gaps between protein sequences and protein tertiary structures. Thus, to fill this gap and overcome this problem, various computational methods have been introduced for the prediction of protein structure (Floudas et al., 2006). One of the methods is known as homology modelling. This modelling method predicts the protein's capacity and coordinates further test work. This method can require less operational time, and the results are as reliable as experimental methods (Marks et al., 2012). In this study, computational protein structure predictions were investigated using the protein chitin deacetylase (Alfonso et al., 1995). This protein was chosen because it has a well-developed crystal structure that was used to compare and validate the 3D model.

METHODOLOGY

Homology modelling is the most common computational method used to predict the 3D structure of a protein. By using the experimentally determined protein structure (template), homology modelling constructs were used to determine the conformation of another protein with a similar amino acid sequence (target) (Tramontano 1998). The precision of forecasting by this technique relies on the sequence similarity (%). If the similarity between the target and template sequence was over 30%, the forecast was considered accurate, and the method was considered as dependable as the experiment method; however, if the sequence identity was under 30% sequence, then the forecast likely contained large errors (Kopp and Schwede 2004). Homology modelling consists of five main steps:

I. Identification of the template sequence. For the target protein chitin deacetylase, the FASTA sequence was downloaded from the PDB [www.rcsb.org] (Berman et al., 2000). The FASTA sequence was used to recognize the template, which is practically identical to the target protein. This step was performed via a BLAST search (Altschul et al., 1997), and at least one suitable template was selected from the results. The selected templates are those sequences that have a high level of comparability and fewer or no missing residues.

II. Alignment of the target and the template sequence. This step involves determining the proportion of residues that are consistent between a pair of protein sequences. The target sequences and the template sequences were aligned together in CLUSTAL OMEGA (Sievers et al., 2011).

III. Generation of the model. After aligning the target and template sequence, the results were saved in PIR format. The 3D model of the protein was generated using MODELLER (Eswar et al., 2007), which uses the method of satisfying spatial restraints.

IV. Assessment of the model. The generated model was assessed to ensure that it was accurate and corresponded to the native structure. The model may present considerable differences from its native structure under certain conditions. One of assessment methods is to compare two protein structures using the root-mean-square-deviation (RMSD) to measure the mean distance between corresponding atoms in the two structures after they have been superimposed. Another method is to use the Structure Analysis and Verification Server (SAVS), which includes the Ramachandran's Plot (Ho and Brasseur 2005) and Verify-3D (Healy et al., 2010) methods.

V. Catalytic site determination. In this research, PyMol (DeLano 2002) was used to view the catalytic site in the protein chitin deacetylase.

RESULTS AND DISCUSSION

Template proteins are very important in developing 3D models by using homology modelling, which compares the target protein and the template protein. In this process, the level of similarity between the target and template protein was measured. From the BLAST results, six template sequences were obtained. Based on the results, the most suitable template for protein chitin deacetylase is protein 2Y8U, which had 42.17% sequence identity similarity. Then, 2Y8U (template protein) was chosen to be aligned together with the protein chitin deacetylase because it had the highest sequence identity among the other templates. Fifty models of the target protein were generated using MODELLER. Every time a bond was formed between two atoms in the protein structure, free energy was released in the form of heat or entropy. The free energy value can determine the stability of the protein structure. In general, the protein structure with the lowest free energy is the most stable. Based on the result obtained from the MODELLER, the lowest free energy was 1468.83 kJ/mol. The generated structure of the protein is shown in Figure 1. The alpha helix presents a ribbon shape, beta strands are shown by arrows, and other components are shown by random coils.

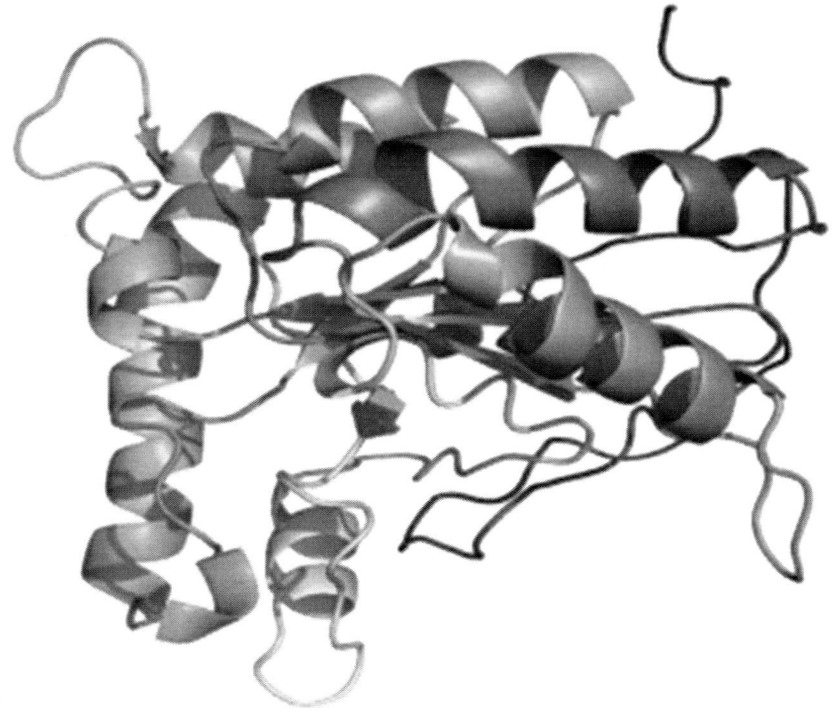

Figure 1. Generated 3D model of chitin deacetylase.

The stereochemical properties of the 3D model of the chitin deacetylase protein were checked by investigating its Ramachandran plot using PROCHECK. Ramachandran's plot was used for basic verification and to determine the conceivable phi (edge around the N-Ca bond) and psi (edge around the Ca-C bond) that record all amino acid residues except glycine and proline. Ramachandran's Plot for the chitin deacetylase protein revealed that among the 247 residues, 91.4% were in the favoured region (red colour region), 8.1% were in the allowed region (yellow colour region) and 0.5% were in the disallowed region (white colour region). For this evaluation technique, a decent quality model presents more than 90% of residues in the favoured district. The second technique of assessing the model is by using Verify 3D, which evaluates the nature of the protein structure by utilizing the scoring capacity. If more than 80% of the deposits have a score of > = 0.2, then the protein structure is considered of high quality. If less than 65% have a score of > = 0.2, then the model is considered to have low quality. The graph of Verify 3d results in Figure 2 demonstrates a similarity score over zero, which indicated that the protein contains favoured side-chain conditions and great crease areas. Although the chart demonstrated that 72.06% of the deposits had normal 3D–1D scores > = 0.2, the model is still acceptable and reliable and has good quality.

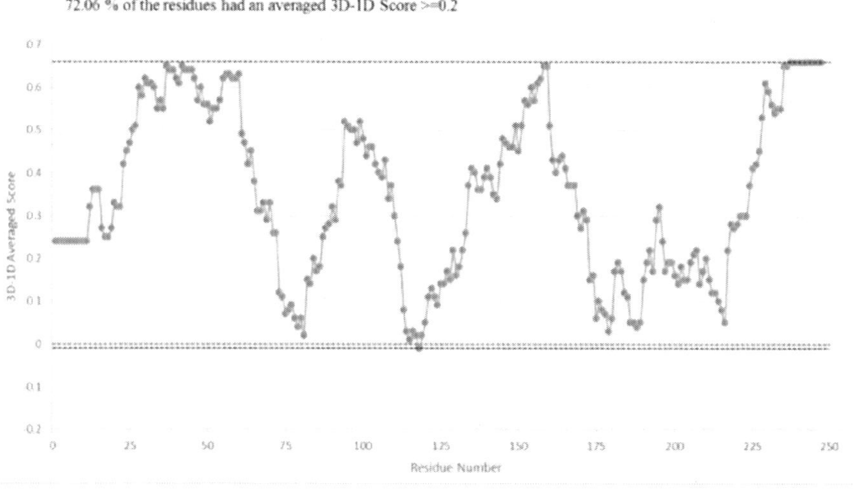

Figure 2. Graph of Verify 3D of chitin deacetylase model.

The third method of evaluating the model is by calculating the RMSD, which is used as a quantitative measure of similarity between the target protein structure (chitin deacetylase) and the template protein structure (2Y8U). The lower the RMSD value, the higher the percentage similarity between the target and template protein structure. Generally, a RMSD value under 3Å would suggest decent similarity between the structures. The RMSD value of the target and template protein structure after superimposition was found to be 1.86 Å, which means that the model was successfully developed.

Finally, the location of the catalytic site residue was determined, which involved the amino acids aspartic acid 41, aspartic acid 42, histidine 100, histidine 104, tyrosine 141 and histidine 205, as shown in Figure 3, where the locations of catalytic site residues were labelled and shown in red. This catalytic site is also comparable to that of the crystal structure.

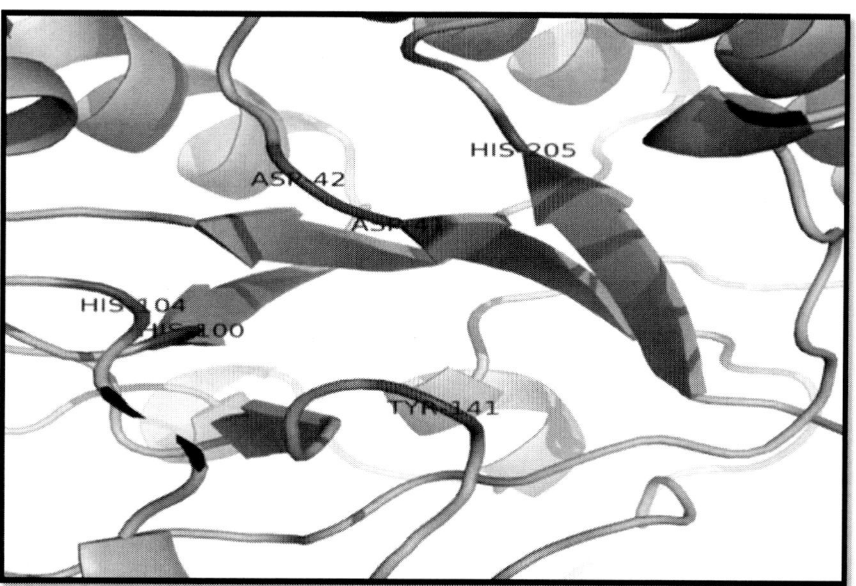

Figure 3. Catalytic site of the chitin deacetylase-generated model.

Conclusion

The predicted protein model closely resembled its native structure. The protein model obtained through homology modelling by the method of satisfying spatial restraints exhibited the most desirable 3D structures relative to the experimentally determined structures. The protein template

(2Y8U), which exhibited high sequence similarity towards the target sequence (42.17%), was also proven to be significant in homology modelling and could obtain a highly accurate model. The computational method of protein structure prediction requires less operational time and has comparable results compared with experimental methods. Furthermore, computational methods are not as costly as experimental methods. Thus, protein structure predictions can be performed as an alternative to experimental methods.

ACKNOWLEDGMENTS

The authors would like to thank Universiti Teknologi Malaysia for providing financial support for this project (GUP Vot No: 19H73).

REFERENCES

Alfonso, C., Nuero, O. M., Santamaría, F. & Reyes, F. (1995). Purification of a heat-stable chitin deacetylase from *Aspergillus nidulans* and its role in cell wall degradation. *Current Microbiology*, 30(1): 49-54.

Altschul, S. F., Madden, T. L., Schaffer, A. A., Zhang, J., Zhang, Z., Miller, W. & Lipman, D. J. (1997). Gapped BLAST and PSI-BLAST: a new generation of protein database search programs. *Nucleic Acids Research*, 25(17): 3389-3402. doi:10.1093/nar/25.17.3389.

Berman, H., Westbrook, J., Feng, Z., Gilliland, G., Bhat, T., Weissig, H. & Bourne, P. (2000). The protein data bank. *Nucleic acids researc*h, 28: 235-242. *URL:* www. rcsb. org Citation.

Carugo, O. & Pongor, S. (2001). A normalized root-mean-spuare distance for comparing protein three-dimensional structures. *Protein Science,* 10(7): 1470-1473.

DeLano, W. L. (2002). *The PyMOL molecular graphics system.*

Eswar, N., Webb, B., Marti-Renom, M. A., Madhusudhan, M., Eramian, D., Shen, M. Y., Pieper, U. & Sali, A. (2007). Comparative protein structure modeling using MODELLER. *Current Protocols in Protein Science*, 50(1): 2.9. 1-2.9. 31.

Floudas, C. A., Fung, H. K., McAllister, S. R., Monnigmann, M. & Rajgaria, R. (2006). Advances in protein structure prediction and de novo protein design: A review. *Chemical Engineering Science,* 61(3): 966-988. doi:10.1016/j.ces.2005.04.009.

Healy, M. B., Athikulwongse, K., Goel, R., Hossain, M. M., Kim, D. H., Lee, Y. J., Lewis, D. L., lin, T. W., liu, C., Jung, M. & Ouellette, B. (2010). Design and analysis of 3D-MAPS: A many-core 3D processor with stacked memory. *Paper presented at the IEEE Custom Integrated Circuits Conference* 2010(pp 1-4). IEEE.

Ho, B. K. & Brasseur, R. (2005). The Ramachandran plots of glycine and pre-proline. *BMC Structural Biology*, 5:14.

Kopp, J. & Schwede, T. (2004). Automated protein structure homology modeling: a progress report. *Pharmacogenomics*, 5(4): 405-416.

Marks, D. S., Hopf, T. A. & Sander, C. (2012). Protein structure prediction from sequence variation. *Nature Biotechnology*, 30(11): 1072.

Sievers, F., Wilm, A., Dineen, D., Gibson, T. J., Karplus, K., Li, W. & Söding, J. (2011). Fast, scalable generation of high-quality protein multiple sequence alignments using Clustal Omega. *Molecular Systems Biology*, 7(1): 539.

Tramontano, A. (1998). Homology modeling with low sequence identity. *Methods*, 14(3): 293-300.

In: The Influence of Ecosystem Services ... ISBN: 978-1-53619-977-2
Editors: Hasmah Abdullah et al.

Chapter 11

EFFECT OF PH ON THE STABILITY OF RECOMBINANT *MYCOBACTERIUM BOVIS* BCG EXPRESSING MSP-1C FROM *PLASMODIUM FALCIPARUM*

Nor Munirah Z., Abdul Aiman Hakim S. N., Abbas M. A. and Rapeah Suppian*

School of Health Sciences, Health Campus Universiti Sains Malaysia, Kubang Kerian, Kelantan, Malaysia

ABSTRACT

A recombinant BCG vaccine expressing the synthetic MSP-1C of *Plasmodium falciparum* was developed in the previous study. The vaccine candidate increased humoral and cellular immune responses in mice, as well as triggering phagocytosis activity and producing pro-inflammatory cytokines that were significantly higher than those produced by the parent BCG clone. However, no research has been done to assess the stability of the vaccine candidate. The acidity or alkalinity of

* Corresponding Author's E-mail: rapeah@usm.my.

the growth medium is one of the most important factors affecting vaccine stability. The purpose of this study was to determine the effect of pH on the stability of the rBCG candidate vaccine. For 14 days, the rBCG culture and parent BCG control were grown in different pH levels of 7H9 broth (pH 5, pH 6, pH 7, and pH 9). On day 14, the viability of the mycobacteria was determined using a spectrophotometer with an optical density (OD) of 600 nm. The effect of pH on the morphology of the mycobacteria was determined using scanning electron microscope while the effect of pH on the stability of MSP-1C gene in the rBCG clone was determined using polymerase chain reaction. The results showed that cloning the MSP-1C gene into the BCG genome does not affect the growth, morphology, or stability of the rBCG vaccine. Both rBCG and the parent BCG control have a similar response to different pH levels of 7H9 broth. The optimum pH for the growth and morphological stability of the rBCG and parent BCG, as well as the stability of the MSP-1C gene in the rBCG clone was at pH 6 - pH 7 respectively, while the growth of the mycobacteria was significantly reduced at pH 5 and pH 9. The results also showed that all the rBCG clones from different pH of 7H9 expressed MSP-1C gene indicating the stability of the gene. Accordingly, these findings paved the way for the clinical application of rBCG as a vaccine against malaria in the future.

Keywords: recombinant BCG, MSP-1C gene, pH, polymerase chain reaction, scanning electron microscope

INTRODUCTION

Malaria is the most crucial life-threatening disease globally and caused 435,000 deaths in 2018 and 445,000 deaths in 2016, with the highest mortality occurring among children under 5 years of age (WHO 2018; WHO 2015). In sub-Saharan Africa, the highest number of cases was observed in children less than five years of age and in pregnant women (Batool, 2015). Human malaria infections are caused by five species of Plasmodium: P. falciparum, P. vivax, P. malariae, P. ovale and P. knowlesi.

The development of an effective vaccine to combat this infection has become an important agenda. At present, strategies are required to control malaria in the early stage, including detection and treatment of malaria infection, and preventive measures aimed at controlling mosquitoes (Cravo et al., 2015). Because of the development of widespread antimalarial drug resistance, treatments have focused on artemisinine-based combination therapy, which is currently the WHO-recommended treatment (Chesne-Seck et al., 2005). Unfortunately, *P. falciparum* resistance to many of the currently used antimalarial drugs has been reported. Resistance to artemisinine derivatives, the most important and widely used antimalarial drug at present, has also been reported (Cravo et al., 2015; Moore et al., 2016).

The spread of drug-resistant malaria parasites and insecticide-resistant mosquitoes has led to a variety of sustained efforts to develop effective malaria vaccine candidates for controlling and eradicating the disease. However, an effective vaccine has not been licenced for malaria (Ouattara and Laurens 2014). Most studies on malaria vaccine development have focused on the asexual blood stage of the parasitic life cycle. At this stage, the merozoites that are released from infected liver cells enter the bloodstream and invade red blood cells (RBCs), which are responsible for all disease symptoms (Paul et al., 2015). Numerous surface proteins of merozoites have been studied as candidate antigens in vaccine development against blood stage malarial infection, including the C-terminus of merozoite surface protein-1 (MSP-1C) (Moss et al., 2012). MSP1 is a leading vaccine candidate at the erythrocytic stage (Guimaraes et al., 2015). Nearly all Plasmodium species infecting humans, simians and rodents contain this molecule (Matsumoto et al., 1998; Birkenmeyer et al., 2010). The Plasmodium yoelii model was the first to demonstrate protective immunity activated by vaccination with MSP1 (Holder and Freeman 1981).

Mycobacterium bovis bacille Calmette–Guérin (BCG), a currently used vaccine for tuberculosis, is one of the most broadly used vectors for the development of recombinant vaccines for various diseases, including malaria. Rigorous studies in mice have shown that the vaccine represents a promising candidate to prevent malaria infection by inducing appropriate humoral and cellular immune responses (Nurul and Norazmi 2011). Previous studies also demonstrated that the vaccine candidate was capable of stimulating higher inflammatory activity in infected macrophages as well as stimulating much higher phagocytic activity, proinflammatory cytokine production and apoptosis activity in mouse and human macrophages compared with BCG and LPS (Rapeah et al., 2010; Mohamad et al., 2014).

The stability of vaccines has a major impact on the success of immunization programs. Instability can reduce the safety and efficacy of candidate antigens in preventing infectious diseases and saving lives. Therefore, determining the stability of vaccine candidates from production to clinical administration is an important part of vaccine development. Significant changes in the stability profile may occur following exposure to environmental stresses, such as temperature and pH, during preparation and handling or because of fluctuations in storage conditions. Therefore, the stability characteristics of each vaccine must be determined empirically to ensure that the minimum standards of potency, identity, and purity continue to be met (WHO 2012). Therefore, this study was conducted to evaluate the importance of pH (one of the factors in evaluating vaccine stability) on the stability of our rBCG candidate vaccine by evaluating the growth and morphology of the mycobacterium in 7H9 growth medium at different pH values (pH 5, pH 6, pH 7, and pH 9) and assess the effect of these media on the stability of the MSP-1C gene in the rBCG clone. In this experiment, parent BCG was used as a control.

METHODOLOGY

Preparation of Mycobacterium Bovis BCG and rBCG Cultures

BCG and rBCG were grown in 7H11 agar supplemented with 10% oleic acid dextrose catalase (OADC) and kanamycin (15 μg/ml for rBCG) and incubated for 2 weeks in an incubator at 37 °C. The cultures were observed for signs of contamination for 2 weeks, and then the growth of colonies was observed since M. bovis BCG is a slow-growing mycobacteria. A single colony of each was aseptically picked, transferred to flasks containing 10 mL of 7H9 broth medium at different pH values (pH 5, pH 6, pH 7 and pH 9), supplemented with OADC and kanamycin and incubated for another 2 weeks in a 37 °C shaker incubator. The pH was adjusted using NaOH or HCl.

Determination of the Growth of BCG and rBCG

The effect of different pH values of 7H9 broth on bacterial growth was determined by measuring the optical density (OD) of the culture at day 0 and day 14 using a spectrophotometer at a wavelength of 600 nm (OD_{600}). Basically, the bacterial suspensions were diluted in 7H9 to obtain a suitable initial OD for the experiments (0.01 to 0.03 at 600 nm) for standardization. One millilitre of the culture was transferred to a specific cuvette for OD measurement. The culture was then incubated at 37 °C for 14 days (2 weeks). On day 14, the OD of the culture was measured again to determine the pattern of bacterial growth.

Determination of the Morphology of BCG and rBCG

The morphological changes (cell shapes) of the BCG and rBCG colonies were determined using scanning electron microscopy (SEM). The BCG and rBCG cultures were centrifuged at 4000 rpm for 15 min. The supernatant was discarded, and the pellet was transferred to a 1.5 ml microcentrifuge tube. McDowel Trump fixative solution was added to the sample, which was further incubated at 4 °C for 2 hours. The sample was washed twice with 1X PBS. Subsequently, 1% osmium tetroxide was added to the sample, which was then incubated for 1 hour at 4°C. Sample processing was continued with a washing step using dH2O twice, which was followed by dehydration with 50% acetone, 75% acetone, 95% acetone and 100% acetone twice for 10 minutes each (except for 100% acetone). Finally, 100% hexamethyldisilazane (HMDS) solution was added to the sample. The sample was then airdried overnight before it was mounted onto a sample tub, coated with gold and viewed under a field-emission scanning electron microscope (FESEM).

Determination of the Stability of the MSP-1C Gene Using Polymerase Chain Reaction (PCR)

The stability of the MSP-1C gene in the rBCG culture was determined using the PCR method. The parent BCG and rBCG cultures were transferred to 15 ml tubes and centrifuged. The supernatant was discarded, and the pellet was washed twice with 1X PBS. The sample was then sonicated for 2 min (15 s on and 15 s off) using an ultrasonic processor XL (sonicator). The sample was then subjected to PCR. BCG was used as a control. Amplification was performed using the specific primers for amplifying MSP-1C, as shown in Table 1.

Table 1. Primers used for PCR analysis

Primer	**Sequence (5' ---------3')**
MSP(M)-MCS(F)	CGGCTAGCATGAACATCAGCCAGCACCAG
MSP(M)-MCS(R)	GCTAGCCCTAGGGTTCGACGACGAGCAGAA

Statistical Analysis

Statistical analyses were carried out using the statistical package of Social Sciences (SPSS), version 20. The data were obtained from three different experiments (n = 3) and are indicated as the means ± standard error of the mean (S.E.M.). The data were analysed by a repeated measure analysis of variance (RM ANOVA), and a value of $^*p < 0.05$ was considered significantly different.

RESULTS

Effect of pH on the Growth of BCG and rBCG Cultures

This study investigated the influence of 7H9 broth with different pH values (pH 5, pH 6, pH 7 and pH 9) on the growth of BCG and rBCG cultures. The optical density (OD) at 600 nm was measured at day 0 and day 14. The results showed that different pH values of 7H9 affected the growth of mycobacteria. Interestingly, both parent BCG and rBCG cultures showed a similar growth pattern. At day 0, the OD of the BCG and rBCG was adjusted to 0.02 to standardize the initial growth. On day 14, BCG and rBCG grew rapidly in 7H9 broth at pH 6 and pH 7 (OD_{600} ~ 0.8-0.9), respectively (Figure 1). However, different results were observed when the pH of the 7H9 broth was adjusted to pH 5 and pH 9. Mycobacteria growth was decreased to OD_{600} ~ 0.5-0.6 when the pH of the medium was adjusted to pH 5, and it also decreased to OD_{600} ~ 0.6-0.7 when the pH of the broth was adjusted to pH 9.

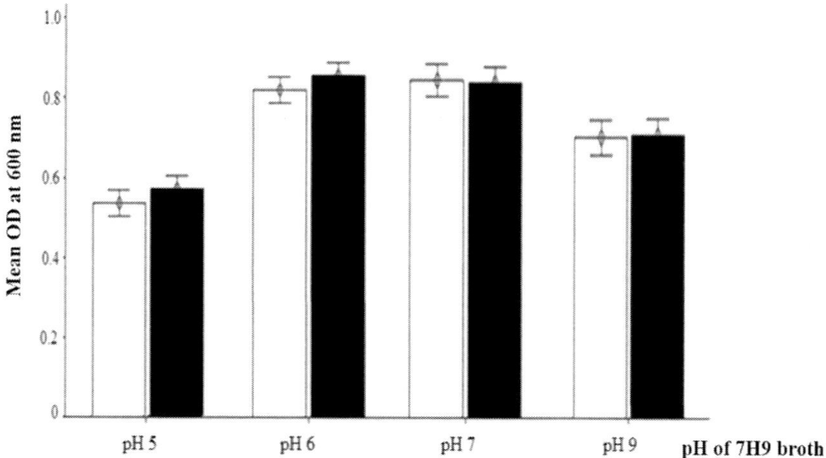

Figure 1. Effect of the pH of 7H9 broth on the growth of BCG and rBCG cultures. Data are presented as the mean OD of mycobacterium at 600 nm ± S.E.M. for three independent experiments, $p < 0.05$.

Effect of pH on Morphology of BCG and rBCG

As shown in Figure 2, alteration of the pH of 7H9 broth affected the cell morphology and staining behaviour of BCG and rBCG after 14 days. In general, morphological differences were not observed between BCG and rBCG isolated from 7H9 broths with different pH values. Under the microscope, both BCG and rBCG were pink and rod-shaped. At pH 5, pH 6 and pH 7, BCG and rBCG cells showed similar morphological cell structures (patterns and sizes). At pH 9, a similar pattern of mycobacteria was observed, although the size of the rod was slightly smaller than that at acidic and neutral pH, which may be related to the increased fragility and partially broken nature of the cells at this pH. Scanning electron microscopy images of the surface indicated that BCG and rBCG did not present differences in structure and morphology. However, under all pH conditions, the rBCG cell had a longer size than the BCG cell. In an acidic environment at pH 5, the cells stick to each other and form clumps or large aggregates, while in an alkaline environment at pH 9, the cells display normal morphology but appear to lose features and are smaller in size.

Effects of pH on the Stability of the MSP-1C Gene in rBCG Cells

Polymerase chain reaction was carried out to determine the effect of different pH values of 7H9 broth on the stability of the MSP-1C gene in the rBCG clone. As shown in Figure 3, the MSP-1C gene was detected at approximately 314 bp in the rBCG cells of all cultures but not in the parent BCG cells, indicating that the gene is still stable. When comparing the intensity of the bands, the band of the MSP-1C gene from rBCG grown in 7H9 broth at pH 5 was significantly lower ($p < 0.05$) than that grown in 7H9 broth at pH 6, pH 7 and pH 9.

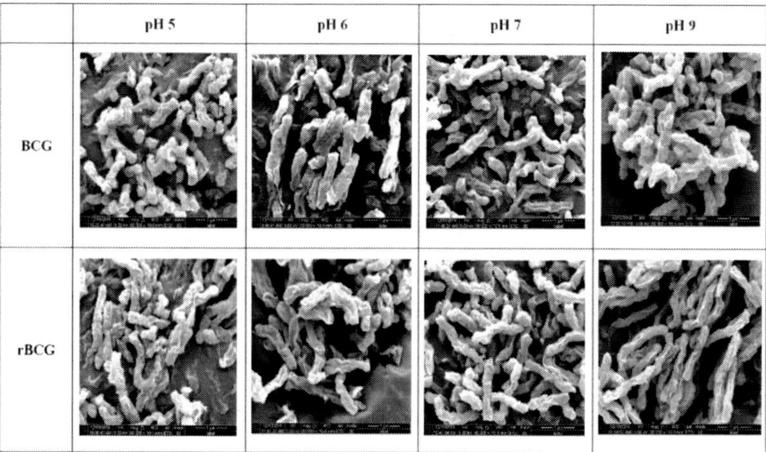

Figure 2. Scanning electron micrographs of BCG and rBCG grown in 7H9 broth with different pH values for 14 days. Photos were captured via scanning electron microscopy (SEM) at a magnification of 60000x.

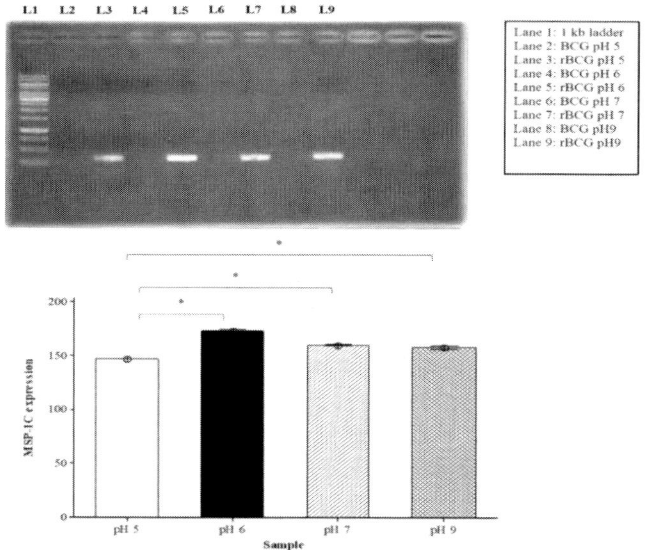

Figure 3. Effect of 7H9 broth pH on the stability of MSP-1C in the rBCG clone. (A) PCR products of the MSP-1C gene of different rBCGs. (B) Comparison of the intensity of MSP-1C bands using ImageJ software. The PCR products were subjected to 1% agarose gel electrophoresis at 80 V for 30 minutes. Data are expressed as the mean relative density of MSP-1C ± S.E.M. of three independent experiments, and $*p < 0.05$ indicated significant differences among different pH values.

The highest band intensity was observed for the MSP-1C gene from rBCG culture grown at pH 6. These data indicated that different pH values of 7H9 broth maintained the presence of the MSP-1C gene but affected the intensity of gene expression in the rBCG clone.

DISCUSSION

Stability is an important factor in the development of a vaccine because it is closely related to the efficacy and safety of the vaccine. Among the factors assessed in testing the stability of a vaccine is the ability of the vaccine to survive in acidic and alkaline conditions. The sensitivity of bacteria in media at different pH values is independent of the type of culture medium and buffer used (Rao et al., 2001). Thus, in this study, we carried out a detailed investigation of how our candidate vaccine, a recombinant BCG vaccine expressing MSP-1C from *P. falciparum*, copes with 7H9 broth growth medium presenting low and high pH values. 7H9 broth was chosen because it is a defined medium for slow-growing mycobacteria, such as *M. bovis* BCG.

Our findings indicate that the growth of BCG and rBCG is inhibited as the external pH decreases to 5 or increases to 9, thus supporting previous findings by Kunisch et al., (2012), who showed that the growth rate of BCG expressing pMV261 in low-pH medium was slightly below its growth rate in neutral medium, as determined by OD measurements. Our results also showed that both BCG and rBCG cells grew well at pH 6 to pH 7, which was expected since pH 6.5 to pH 7 is the optimal pH for 7H9 broth used in normal BCG culture (Portaels and Pattyn 1982). These data are supported by a previous statement suggesting that slow-growing mycobacteria, such as *M. bovis* BCG, cannot grow over a wider pH range (Portaels and Pattyn 1982) and present inhibited growth under extreme pH (Bakermans and Nealson 2004; Nascimento et al., 2004). In general, at acidic pH, the growth of most slow-growing mycobacteria was less than that at neutral pH. Extreme pH has adverse effects on cell morphology and cell lysis (Sabri and Hasnain 1994). Lysis may be associated with the effects of biochemical reactions, membrane proteins, membrane lipids and septation (Kubitschek 1990; Intriago and FloodGate 1991) or changes in the cell wall components, peptidoglycan contents and cell envelope defects (Shohayeb and Chopra 1987; Waximan and Strominger 1983). This hypothesis is consistent with our observation showing that the pellets of BCG and rBCG grown in 7H9 broth adjusted to pH 5 were lysed and difficult to isolate.

The effect of mycobacteria on acidic environments has been studied extensively in a previous study. Vandal et al., (2009) showed that certain mycobacteria, such as *M. tuberculosis* and *M. smegmatis*, were able to cope with the stress of acidic environments using acid resistance mechanisms. In this study, both BCG and rBCG could still grow in an acidic environment, suggesting that the bacteria have an acid resistance mechanism, however, the growth rate was reduced compared to those grown at other pH values of 7H9. This phenomenon might be due to the structure of the cell wall of the bacteria. The physical structure and molecular composition of bacterial cell envelopes act as an effective primary barrier against the entry of protons. If protons enter the bacterial cytosol, an array of mechanisms are activated to regulate the pH within the cell to maintain the intrabacterial pH and promote the survival of bacteria in acidic environments (reviewed in Vandal et al., 2009).

Our data also showed that both BCG and rBCG can grow under alkaline conditions at pH 9 and presented much better growth than at pH 5. This result indicates that mycobacteria have an alkaline stress mechanism. However, acid fast staining and scanning electron microscopic analyses demonstrated that growing the mycobacteria in 7H9 broth at pH 9 reduced the morphological strength of the cell. This condition might be due to the presence of OH- ions in alkaline medium that saponify the lipids in the enveloping membrane of the mycobacteria, thus leading to the destruction of its structure. A higher pH of the medium or solution disorganizes the structure of the peptidoglycane and causes hydrolysis of the nucleotides of the mycobacteria (Maris 1992).

The PCR results showed that the MSP-1C gene was present in the rBCG genome for the bacteria were grown under different pH conditions, indicating that different pH values did not substantially affect the stability of the MSP-1C gene. However, the rBCG cells exhibited higher expression of the MSP-1C gene at pH 6 than at other pH values, which strongly supports the data from the growth and morphological observations. This result is in agreement with a previous study in which the the propanediol utilization (Pdu) bacterial microcompartments (MCP) shell of *Salmonella enterica* was found to be stable in buffers with pH 6 – pH 10 (Kim et al., 2014). In another study, Gupta et al., (1995) evaluated the effect of different pH values on the stability of a recombinant shuttle plasmid, pCPPS-31, in *Escherichia coli*. Similar to our results, their findings showed that pH values ranging from 5-8 had no significant effect on plasmid stability. The stability of the MSP-1C gene in the rBCG genome is important for ensuring that the protein expressed by BCG is the expected protein and can stimulate a specific immune response against the malarial parasite. The structural stability of proteins is of crucial importance for the efficient presentation of antigenic peptides on MHCs, which play a decisive role in triggering strong immune reactions. Any changes to the gene may lead to the expression of a protein that fails to stimulate a specific immune response or becomes toxic (Scheiblhofer et al., 2017).

CONCLUSION

In conclusion, the results showed that different pH values of 7H9 broth have a significant effect on the growth of parent BCG and rBCG. Similar responses to different pH values were observed in the parent BCG and rBCG. The optimal pH for the growth and stability of BCG and rBCG was pH 6 – pH 7. The growth of the mycobacteria was significantly reduced at pH 5 and pH 9, while at pH 6 – pH 7, the mycobacteria showed high growth rates. Acidic conditions also affected the behaviour of mycobacteria, with the cells becoming aggregates and difficult to isolate. Moreover, all the rBCG clones from 7H9 with different pH values expressed the MSP-1C gene, indicating the stability of the gene in the rBCG clone. However, higher expression was observed in rBCG grown in

7H9 at pH 6. The findings of this study support the use of rBCG as a vaccine candidate against malaria in the future.

ACKNOWLEDGMENTS

This research was supported by Universiti Sains Malaysia (USM) Research University Grants (RUI) (1001/PPSK/8011100).

CONFLICT OF INTEREST

The authors declare that they have no conflicts of interest.

REFERENCES

Bakermans, C. & Nealson, K. H. (2004). Relationship of critical temperature to macromolecular synthesis and growth yield in *Psychrobacter cryopegella. Journal of Bacteriology*, 186(8): 2340-2345.

Batool, S. M. (2015). Malaria in pregnant women. *International Journal of Infection*, 2(3): e22992.

Birkenmeyer, L., Muerhoff, A. S., Dawson, G. J. & Desai, S. M. (2010). Isolation and characterization of the MSP1 genes from *Plasmodium malariae* and *Plasmodium ovale. The American Journal of Tropical Medicine and Hygiene*, 82(6): 996-1003.

Chesne-Seck, M.-L., Pizarro, J. C., Vulliez-Le Normand, B., Collins, C. R., Blackman, M. J., Faber, B. W., Remarque, E. J., Kocken, C. H., Thomas, A. W. & Bentley, G. A. (2005). Structural comparison of apical membrane antigen 1 orthologues and paralogues in apicomplexan parasites. *Molecular and Biochemical Parasitology*, 144(1): 55-67.

Cravo, P., Napolitano, H. & Culleton, R. (2015). How genomics is contributing to the fight against artemisinin-resistant malaria parasites. *Acta Tropica*, 148:1-7.

Guimarães, L. O., Wunderlich, G., Alves, J. M., Bueno, M. G., Röhe, F., Catão-Dias, J. L., Neves, A., Malafronte, R. S., Curado, I. & Domingues, W. (2015). Merozoite surface protein-1 genetic diversity in *Plasmodium malariae* and *Plasmodium brasilianum* from Brazil. *BMC infectious diseases*, 15(1): 529.

Gupta, R., Sharma, P. & Vyas, V.V. (1995). Effect of growth environment on the stability of a recombinant shuttle plasmid, pCPPS-31, in *Escherichia coli*. *Journal of Biotechnology*, 41(1):29-37.

Holder, A. A. & Freeman, R. R. (1981). Immunization against blood-stage rodent malaria using purified parasite antigens. *Nature*, 294(5839): 361-364.

Intriago P. & Floodgate G.D. (1991). Fatty acid composition of the estuarine *Flexibacter* sp. strain Inp: effect of salinity, temperature and carbon source for growth. *Journal of General Microbiology*, 137:1503–1509.

Kim, E. Y., Marilyn F Slininger, M. F. & Tullman-Ercek, D. (2014). The effects of time, temperature, and pH on the stability of PDU bacterial microcompartments. *Protein Science*, 23(10): 1434-1441.

Kubitschek, H. E. (1990). Cell volume increase in *Escherichia coli* after shifts to richer media. *Journal of Bacteriology*, 172: 94-101.

Kunisch, R., Kamal, E., & Astrid Lewin, A. (2012). The role of the mycobacterial DNA-binding protein 1 (MDP1) from *Mycobacterium bovis* BCG in host cell interaction. *BMC Microbiology,* 12:165.

Maris, P. (1995). Modes of action of disinfectants. *Scientific and Technical Review of the Office International des Epizooties,* 14(1): 47-55.

Matsumoto, S., Yukitake, H., Kanbara, H. & Yamada, T. (1998). Recombinant *Mycobacterium bovis* bacillus Calmette-Guerin secreting merozoite surface protein 1 (MSP1) induces protection against rodent malaria parasite infection depending on MSP1-stimulated interferon-γ and parasite-specific antibodies. *Journal of Experimental Medicine*, 188(5): 845-854.

Mohamad, D., Suppian, R. & Mohd Nor, N. (2014). Immunomodulatory effects of recombinant BCG expressing MSP-1C of *Plasmodium falciparum* on LPS-or LPS+ IFN-γ-stimulated J774A. 1 cells. *Human Vaccines & Immunotherapeutics*, 10(7): 1880-1886.

Moore, K. A., Simpson, J. A., Paw, M. K., Pimanpanarak, M., Wiladphaingern, J., Rijken, M. J., Jittamala, P., White, N. J., Fowkes, F. J. & Nosten, F. (2016). Safety of artemisinins in first trimester of prospectively followed pregnancies: an observational study. *The Lancet Infectious Diseases*, 16(5): 576-583.

Moss, D. K., Remarque, E. J., Faber, B. W., Cavanagh, D. R., Arnot, D. E., Thomas, A. W. & Holder, A. A. (2012). *Plasmodium falciparum* 19-kilodalton merozoite surface protein 1 (MSP1)-specific antibodies that interfere with parasite growth *in vitro* can inhibit MSP1 processing, merozoite invasion, and intracellular parasite development. *Infection and Immunity*, 80(3): 1280-1287.

Nascimento, M. M., Lemos, J. A, Abranches, J., Gonçalves, R. B. & Burne R. A. (2004). Adaptive acid tolerance response of *Streptococcus sobrinus*. *Journal of Bacteriology*, 186(19):6383–6390.

Nurul, A. A. & Norazmi, M. N. (2011). Immunogenicity and in vitro protective efficacy of recombinant *Mycobacterium bovis* bacille Calmette Guerin (rBCG) expressing the 19 kDa merozoite surface protein-1 (MSP-119) antigen of *Plasmodium falciparum*. *Parasitology Research*, 108(4): 887-897.

Ouattara, A. & Laurens, M. B. (2014). Vaccines against malaria. *Clinical Infectious Diseases*, 60(6): 930-936.

Paul, A. S., Egan, E. S. & Duraisingh, M. T. (2015). Host-parasite interactions that guide red blood cell invasion by malaria parasites. *Current Opinion in Haematology*, 22(3): 220.

Portaels, F. & Pattyn, S. R. (1982). Growth of mycobacteria in relation to the pH of the medium. *Annals of Microbiology*, 133(2): 213-221.

Rao, M., Streur, T. L., Aldwell, F. E. & Gregory M. (2001). Cook intracellular pH regulation by *Mycobacterium smegmatis* and *Mycobacterium bovis* BCG. *Microbiology*, 147:1017–1024.

Rapeah, S., Dhaniah, M., Nurul, A. A. & Norazmi, M. N. (2010). Phagocytic activity and pro-inflammatory cytokines production by the murine macrophage cell line J774A. 1 stimulated by a recombinant BCG (rBCG) expressing the MSP1-C of *Plasmodium falciparum*. *Tropical Medicine*, 27(3): 461-469.

Sabri, A. N. & S. Hasnain. (1996). Polypeptide products in diviva+ and diviva-strains of *Bacillus subtilis* under varying temperature, pH and stavation conditions. *Pakistan Journal of Zoology*, 28: 303-309.

Scheiblhofer, S., Laimer, J., Machado, Y., Weiss, R. & Josef Thalhamer, J. (2017). Influence of protein fold stability on immunogenicity and its implications for vaccine design. *Expert Review of Vaccines*, 16(5): 479–489.

Shohayeb, M. & Chopra, I (1987). Mutations affecting penicillin-binding proteins 2a, 2b and 3 in *Bacillus subtilis* alter cell shape and peptidoglycan metabolism. *Journal of General Microbiology*, 133: 1733-1742.

Vandal, O. H., Nathan, C. F. & Ehrt, S. (2009). Acid resistance in *Mycobacterium tuberculosis*. *Journal of Bacteriology,* 191:4714-4721.

Waxman, D. J. & Strominger, J. L. (1983). Penicillin-binding proteins and the mechanism of action of Beta-Lactam antibiotics· *Annual Review of Biochemistry,* 52: 825–869.

World Health Organization (WHO). (2012). *World malaria report.* Geneva 2012.

World Health Organization and Maternal and Child Health Epidemiology and Estimation Group. (2016). *World Health Organization.* (2015). World malaria report. Geneva, 2015.

ABOUT THE EDITORS

Hasmah Abdullah is an Associate Professor at Universiti Sains Malaysia, Health Campus, Kubang Kerian, Malaysia. She is a PhD holder in Biochemistry. Her background in Biology triggered the interest in writing issues pertaining to environment and health. Her current research focuses on the application of biochemistry to elucidate the mechanism of action of natural products towards human cancer cells.

Rapeah Suppian is an Associate Professor at Universiti Sains Malaysia, Health Campus, Kubang Kerian, Malaysia. She is a PhD holder in Immunology. As an academician, she is concerned about the environment and health issues pertaining to ecosystem services which affect the sustainability of natural resources. Her current research focuses on the immunoregulatory effect of natural products towards immune cells such as macrophages.

INDEX

A

B

C

I

K

L

M

N

O

P

Q

R

S

T

U

V

W

Z